ALGEBRA

This book provides an i̱   foundations of Algebra, with p   ̱ appⅼications to
engineering and allied s̱   ̱. ⅼhe progressive exercises are
designed both to test the reader's understanding of the subject and
to give practice in the essential power of manipulation. Only a
knowledge of arithmetic is assumed, though some reference is
made to theorems in geometry and trigonometry for the benefit
of those students who are acquainted with them.

23 NOVEMBER
TUESDAY
LECTURE
THEATRE 2

RICHARD C. MCKAY.

TEACH YOURSELF BOOKS

Granted that it is possible to learn the subject from a book, then this volume will serve the purpose excellently. It covers the ground from the very beginning, along the usual paths of equations, factors, indices and so on, to a concluding chapter on simple progressions and an appendix which contains notes on Permutations, the Binomial Theorem and Quadratic Equations theory. A chapter worthy of especial mention is that dealing with Determination of Graph Laws, a topic too often neglected but one of great value and frequent practical use.

Regarded as a textbook, this is probably the best value for money on the market.

Higher Education Journal

ALGEBRA

P. Abbott
and
M. E. Wardle

TEACH YOURSELF BOOKS

Hodder and Stoughton

First printed 1942
Revised edition 1971
Fourth impression 1993

Copyright © 1971 edition
Hodder and Stoughton Ltd
Copyright © 1991 edition
Michael Wardle

A CIP catalogue record for this title is available from the British Library.

ISBN 0–340–54914–9

Printed in Great Britain
for Hodder and Stoughton Educational, a division of Hodder
and Stoughton Ltd, Mill Road, Dunton Green, Sevenoaks,
Kent, by Clays Ltd, St Ives plc
Typeset by Taurus Graphics, Abingdon, Oxon.

Contents

Introduction

Algebra is such a wide and comprehensive subject that this volume cannot be regarded as anything more than an elementary introduction to it. It is an endeavour to enable the private student to learn something of the principles and foundations of the subject, thus enabling you to proceed to the study of more detailed and advanced treatises. It also provides, within the necessarily prescribed limits of such a book, that knowledge of algebra which is required by a student of allied branches of mathematics or in applications of mathematics to engineering, etc. Consequently some of those elementary sections of the subject which are of little use for these purposes have not been included.

The exercises are progressive and designed both to enable you to test your knowledge of the work you have studied and also to provide material for your training in that power of manipulation which is so essential. They contain few of the more complicated or academic problems which are beyond the practical requirements of the ordinary student.

An appendix contains, without exercises, a very brief summary of the meaning of permutations and combinations, the binomial theorem, and the nature of the roots of a quadratic equation, together with an introduction to set notation and the laws of set algebra. Also included are those formulae which you may require when beginning work on the calculus or other branches of mathematics.

While the fundamental laws of algebra have not been entirely overlooked, rigid proofs of them have been omitted, owing to exigencies of space. It is hoped, however, that the logical basis of the subject has not been seriously impaired by the omissions.

Some emphasis has been placed on the graphical aspects of parts of the subject, since experience has shown that they prove stimulating and provide revealing help to the student.

No previous mathematical knowledge is required for this work, beyond that of arithmetic. References have occasionally been made to theorems in geometry or trigonometry for the benefit of those students who have some knowledge of them.

The authors would like to thank Mr C. E. Kerridge for permission to use examples from *National Certificate Mathematics* Vol I and Mr H. Marshall for the use of examples from Vol II.

1

The Meaning of Algebra

1 Algebra and Arithmetic

Algebra, like arithmetic, deals with numbers. Fundamentally the two subjects have much in common; indeed, algebra has been called 'generalised arithmetic', though this is a very incomplete description of it. It would perhaps be more correct to say that algebra is an extension of arithmetic.

Both subjects employ the fundamental operations of addition, subtraction, multiplication and division of numbers, subject to the same laws. In each the same symbols, $+$, $-$, $\times$, $\div$, are used to indicate these operations, but in algebra, as new processes are developed, new symbols are invented to assist the operations. Terms such as fractions, ratio, proportion, square root, etc., have the same meaning in both subjects, and the same rules govern their use.

In arithmetic we employ **definite numbers**; we operate with these and obtain definite numerical results. Whereas in algebra, while we may use definite numbers on occasions, we are, in the main, concerned with **general expression** and **general results**, in which letters or other symbols represent numbers not named or specified.

This may seem vague to a beginner, but the following simple example may serve to show what is meant.

2 A formula

In arithmetic we learn that to find the area of the floor of a room, rectangular in shape, we 'multiply the length of the room by the breadth'. This might be expressed in the form:

The area of a rectangle in square metres is equal to the length in metres multiplied by the breadth in metres.

This rule is shortened in algebra by employing letters as symbols, to represent the quantities. Thus:

Let the letter l represent 'the length in metres'.
Let the letter b represent 'the breadth in metres'.
Let the letter A represent 'the area in square metres'.

With these symbols the above rule can now be written in the form:

$$A = l \times b$$

In this shortened form we express the rule for finding the area of any rectangle; it is a general rule, and is called a formula. The full description of it would be

'The formula for finding the area of a rectangle.'

It will be noticed that in the formula above there is no mention of units. This is because it is true whatever units are employed. It is necessary, however, that the same kind of unit should be employed throughout. If l and b are measured in mm, A will be area in square mm; if they are measured in m, A will be in square m.

When using the formula for any specific case it is important to state clearly what unit is employed for l and b. The unit for A will then follow.

Algebra has been called a kind of shorthand, and the above example in which a sentence has been reduced to

$$A = l \times b$$

illustrates the reason for the description, but as progress is made you will discover that it is an incomplete description. From such a simple beginning as the above the subject develops into the most powerful instrument employed in mathematics.

The Greeks produced some of the greatest mathematicians in history, but their work was mainly accomplished in geometry. They made very little progress in algebra, since the symbols they could employ were extremely few. They did not even possess separate symbols for numbers, such as were afterwards introduced through the Arabs, and which we use today, but used instead the letters of the alphabet to represent numbers. The Romans were similarly restricted, and neither nation employed the decimal system of notation. You will realise something of the great difference suitable symbols make in mathematics if, as an example, you write down the number of the year in Roman numerals and then try to multiply it, say by 18, also expressed in Roman fashion.

For progress in manipulation and general development of the subject not only do we require symbols, but they must be suitable for the purpose. The history of mathematics reveals that many of the symbols which are so familiar to us now have been reached by a slow evolutionary process, often lasting through centuries.

When letters are used to represent numbers any suitable choice may be made. In the above formula, although A, l and b were employed, any other letters could be used if they were more suitable. By common usage, however, certain letters are usually selected for specific purposes, and the same symbols are used invariably to denote certain numbers, as will become apparent to you as you progress.

But whatever letters are employed in solving problems, you must carefully observe the following:

(1) It must be clearly stated what each symbol represents.
(2) If measurements of any kind are involved, the unit employed must be clearly defined.

3 Transformation of a formula

The student may be inclined to wonder why such an elementary rule as the area of a rectangle should be expressed as a formula by the employment of algebraic symbols. But this is an example selected for its simplicity to illustrate the meaning of an algebraic formula. Even in this case it is possible in some measure to illustrate the flexibility and adaptability of a formula as compared with a statement in words.

For example, suppose that l and b were measured in cm, then A = l × b square centimetres.

If it is required to express this result in square metres we could do it thus:

$$A = \frac{l \times b}{10\ 000} \text{ square metres}$$

Suppose that the rectangle is a room, l and b being measured in cm, and that the room is covered with carpet costing a pence per square metre. If the cost of the whole carpet be represented by C pence, then:

Since
$$A = \frac{l \times b}{10\ 000} \text{ square metres}$$

then
$$C = \pounds\, \frac{l \times b}{10\ 000} \times a \text{ pence}$$

If this were to be expressed in pounds, then:

$$C = \pounds\, \frac{l \times b}{10\ 000} \times \frac{a}{100}.$$

Thus a formula may be used as a foundation for other formulae to express modifications of the original.

There is a convenient notation which is employed throughout mathematics and it is more fully discussed in Section 18. Briefly, when a number is multiplied by itself, this is called squaring the number: thus, $a \times a$ is squared and is written a^2. Similarly, 1 square metre, the unit of area which is $1\ m \times 1\ m$, is written as $1\ m^2$. 28 square metres would be written $28\ m^2$. Again, m^3 is $m \times m \times m$ and represents cubic metres.

4 Examples of using formulae

Foreign currency

When travelling abroad it is usually necessary to change the money of your own country into that of the country that you are visiting. A formula is used to convert one currency into the other. In order to do this you need to know the current exchange rate. For example, in October 1990, the number of German marks you might get in exchange for £1 was 2.95. This can be shown as the formula: *Number of marks is number of pounds times 2.95*, or $M = 2.95 \times S$, where M is used to represent the number of marks and S is used to represent the number of pounds sterling. To find the number of marks you would get for £50 you replace the letter S in the formula by 50. So the number of marks you would get is 2.95×50 i.e. 147.5 marks. If you wanted to change marks into pounds then you would have to rewrite the formula as $S = M/2.95$. To find the number of pounds you have to divide the number of marks by 2.95.

Income Tax

Income tax is based on your income after certain allowances have been deducted. In 1990 a single person was allowed to earn at least £3005 before paying any tax. To work out the tax due on an income of £11425, you first subtract the allowance of £3005 from the income of £11425, and then find 25% of the figure remaining, i.e. 25% of (11425 − 3005). In this case the tax due is 25% of £8420 which is £2105. This can be shown as the formula: *Tax due is 25% of (income minus allowances)* or T = 25% of (I − A), where T is used to represent the tax due, I is the total income and A are the deductable allowances.

Car hire

Often when hiring a car you are charged a standard amount plus so much for each mile. For example the charge might be £12 plus 15p for each mile that you use the car. In this case the charge for a total journey of 80 miles would be £12 + 80 × 15p, i.e. £24. This can be shown as the formula: *Total cost is standard charge plus number of miles times 15p* or C = S + M × 0.15, where C is the total charge, assumed to be in pounds, S is the standard charge and M is the number of miles travelled. Note that it was necessary to write the 15p as 0.15 in the formula so that the units of each of the terms are the same.

5 An illustration from numbers

The following example illustrates the use of letters to represent generalisations in number. We know that:

(1) If any integer – i.e., a whole number – be multiplied by 2, the result is always an even number.
(2) If any even number be increased by unity, the result is an odd number.

These two statements can be combined in one as follows:

If any integer be multiplied by two and the product increased by unity, the result is an odd number.

This is a generalisation about an odd number, expressed in words. This can be expressed by means of algebraical symbols as follows:

(1) Let n be any integer.

(2) Then $2 \times n$ is always an even number

(3) Therefore $2 \times n + 1$ is always an odd number.

In (3) we have reached an algebraical expression by means of which any odd number can be represented.

The brevity and lucidity of this expression as compared with the full description of an odd number above will be apparent. But its value goes beyond this. We can manipulate this algebraic form, we can operate with it, and so can use it in the solution of problems.

We may first note, however, that when expressing the product of two or more numbers represented by letters or a numeral and letters, the sign of multiplication can be omitted. Thus $2 \times n$ can be written as $2n$, and $2 \times n + 1$ as $2n + 1$. This cannot be done with two numerals, such as 25, because under the decimal system the figure 2 has a place value. Multiplication may also be shown by a dot, thus $2.n$.

If any odd number can be represented by $2n + 1$, then, since when any odd or even number is increased by 2 the result is the next odd or even number, therefore the next odd number greater than $2n + 1$ is $2n + 1 + 2$ or $2n + 3$.

Similarly $2n + 5$ is the next odd number above $2n + 3$.

Consequently the expressions

$$2n + 1, 2n + 3, 2n + 5, 2n + 7 \ldots \text{etc.,}$$

represent a series, or a succession of consecutive increasing odd numbers.

Similarly, if an even number be diminished by 1, we obtain an odd number.

$2n - 1, 2n - 3, 2n - 5 \ldots$, represent a series of decreasing odd numbers.

Note: The succession of 'dots' after the sets of odd and even numbers indicates that we could write down more such numbers if it were necessary.

6 Substitution

In the algebraic representation of a set of odd numbers – viz.:

$$2n + 1, 2n + 3, 2n + 5 \ldots$$

since n represents any integer, we could, by assigning some particular value to n, obtain the corresponding odd number.

Thus if $n = 50$

Then
$$2n + 1 = 2 \times 50 + 1$$
$$= 100 + 1$$
$$= 101.$$

Similarly,
$$2n + 3 = 2 \times 50 + 3$$
$$= 100 + 3$$
$$= 103.$$

Consequently the series of increasing odd numbers corresponding to this particular value of n is:

$$101, 103, 105, 107 \ldots$$

Similarly, the decreasing odd numbers when $n = 50$ can be found by substituting this value of n in the series:

$$2n - 1, 2n - 3, 2n - 5 \ldots$$

Then we get the corresponding arithmetical values:

$$99, 97, 95 \ldots$$

7   Examples of generalising patterns

There are many situations which can be described using an algebraic expression or formula. The next three examples show how a typical pattern in a sequence can be described using letters.

Look at the pattern of dots in each of the shapes below.

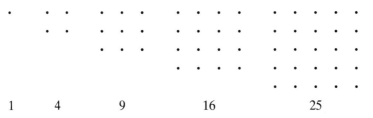

1   4   9   16   25

In the second pattern there are $2 \times 2(= 4)$ dots, in the third pattern there are $3 \times 3(= 9)$ dots, in the fourth pattern there are $4 \times 4(= 16)$ dots. In the next pattern there are $5 \times 5(= 25)$ dots. If the pattern continued in the same way there would be 10×10 $(= 100)$ dots in the tenth shape. To find the number of dots in any particular pattern you find the position of the pattern in the sequence and then multiply this number by itself. The number of dots in the seventeenth pattern in the sequence would be 17×17

(= 289) dots. In a similar way in the nth pattern there would be n × n (or n²) dots.

Look at the pattern of shapes below. Each one is made up of a different number of lines. The first has 3 lines, the second has 5 lines, the third has 7 lines, the next again has two more lines and so on. These are the odd numbers. Each number of lines is two more than the previous number.

The number of lines in any given shape can be related to the position of the shape in the sequence. The fourth shape has 2 × 4 + 1(= 9) lines, the fifth shape has 2 × 5 + 1(= 11) lines, the sixth shape has 2 × 6 + 1(= 13) lines, whilst the tenth shape would have 2 × 10 + 1(= 21) lines.

To find the number of lines in any particular shape you find the position of the shape in the sequence, double this number and then add one.

Letters can be used to describe the number of lines in a particular shape. If we use the letter n to describe the position of the nth shape in the sequence, then the number of lines can be written as 2 × n + 1, or 2n + 1. This expression can be used to write down the number of lines in any shape. For example in the 17th shape the value of n to be substituted in 2n + 1 is 17, so the number of lines is 2 × 17 + 1(= 35) lines.

Look at the second pattern of shapes below. This time each one is made up of a number of squares. As in the previous example we can find an expression which allows us to write down the number of lines in any given shape.

The number of lines increases by three for each new shape. The number of lines in any given shape can be related to its position in the sequence. The first shape has 1 + 3(= 4) lines, the second shape has 1 + 3 + 3(= 7) lines, the third shape has 1 + 3 + 3 + 3 (= 10) lines, whilst the tenth shape would have 1 + 10 × 3(= 31) lines.

To find the number of lines in any particular shape you find the number of squares in the shape, multiply this number by 3 and then add 1.

Letters can be used to describe the number of lines in a particular shape. If we use the letter n to describe the number of squares in the nth shape in the sequence, then the number of lines can be written as $3 \times n + 1$, or $3n + 1$. This expression can be used to write down the number of lines in any shape. For example in the 17th shape the value of n to be substituted in $3n + 1$ is 17, so the number of lines is $3 \times 17 + 1(= 52)$ lines.

There are various ways of looking at the number of lines in the pattern of squares above. Each of these can lead to a different way of describing the typical shape. For example in each shape the number of horizontal lines is double the number of squares in the shape and the number of vertical lines is one more than the number of squares. If we use the letter n to represent the number of squares in the shape then the total number of lines is $2n + (n + 1)$ which can also be written as $3n + 1$ as before. Another way might be to take the number of lines in each of the n squares and to subtract the number of vertical lines where two adjacent squares touch, since these have been counted twice. The number of lines in n squares is $4n$. The number of lines counted twice is one less than the number of squares i.e. $n - 1$, so the total number of lines in the n squares is $4n - (n - 1)$ which can also be rewritten as $3n + 1$ as before.

Using letters to describe the general case, in situations like these, is a very powerful way of summarising what might otherwise be an extremely lengthy process. Algebraic expressions and formulae are the mathematician's shorthand which in many cases allow information to be both concise and extremely precise at one and the same time.

8 Letters represent numbers, not quantities

Such things as length, weight, cost, are called quantities. In short, anything which can be measured is called a quantity.

Letters are not used to represent these quantities. Thus in the formula which we stated above – viz.:

$$A = l \times b$$

l stands for the **number** of cm or m, as the case may be, and to obtain it we must first determine the unit to measure the quantity, and then use a letter to represent the number of units.

It is important, therefore, that when letters are to be used in algebraical expressions, it should be clearly stated what each letter represents.

Thus we should say:

Let l represent the number of cm in the length or, more briefly, let the length be l cm.

Similarly let n represent the number of men, and let c represent the cost in pence.

We do sometimes, for brevity, and somwhat loosely, speak, for example, of the area of a rectangle as A, when we should say A m^2 or A units of area.

The sign = may be used loosely in abbreviation of the statements above. Thus we could write:

Let l = the length in metres,

or let c = the cost in pence.

The sign = means 'equals' or 'is equal to'. It should connect two expressions which are equal in magnitude, but is often used loosely as above to express equality.

9 Examples of algebraic forms

We now give a few examples of what may be termed algebraic forms – i.e., the expression in algebraic symbols and signs of operation of statements about quantities.

Example 1. Express in algebraic form the number of pence in x pounds added to y pence.

To express pounds in pence we multiply by 100.

x pounds = 100 × x pence,

and the total number of pence is 100x + y.

Example 2. A car travels for t hours at v km/h. How far does it go? How far will it go in 20 min?

The car goes v km in 1 h and so 2v km in 2 h, 3v km in 3 h and so on; in t h it will go t × v km.

20 min is $\frac{1}{3}$ h and so the car travels $\frac{1}{3}$ × v km in 20 min.

In $\frac{1}{3}$ h the car travels $\frac{v}{3}$ kilometres.

Example 3. There are two numbers; the first is multiplied by 3 and 5 is added to the product. This sum is divided by 4 times the second number. Express the result in algebraic form.

We must begin by choosing letters to represent the unknown number.

Let x represent the first number.

Let y represent the second number.

Then three times x increased by 5 will be represented by 3x + 5, and four times the second number is 4y.

The division of the first expression by this is $\dfrac{3x + 5}{4y}$.

Exercise 1

1 Write down expressions for:
   (1) The number of pence in £x.
   (2) The number of pounds in n pence.
2 If £a be divided among n boys, how many pence will each boy get?
3 If n men subscribe £a each and m other men subscribe b pence each, how many pence are subscribed in all?
4 Write down the number of
   (1) Metres in a kilometres.
   (2) Men in x Apollo spacecraft.
   (3) Tonnes in y kilogrammes.
5 The sum of two numbers is 28. If one number be n, what is the other?
6 The difference between two numbers is x; if one of them is 50, what is the other?
7 The product of two numbers is a and one of them is x; what is the other?
8 If the average length of a man's step is x cm,
   (1) How far will he walk in 100 steps?
   (2) How many steps will he take in walking 1.8 km?
9 A number (n) is multiplied by 2, and 5 is added to the sum. Write down an expression for the result.
10 If x is an odd number write down expressions for
   (1) The next odd numbers above and below it.
   (2) The next even numbers above and below it.
11 A man buys sheep; x of them cost him a pence each and y of them cost b pence each. What was the total cost in pounds?
12 A number is represented by x; double it, add 5 to the result, and then divide the whole by 6y. Write down an expression for the result.
13 What number must be subtracted from a to give b?
14 What number divided by x gives y as a quotient?

15   What is the number which exceeds b by a?

16   The numerator of a fraction is x increased by 2. The denominator is y diminished by 5. Write down the fraction.

17   A car travels for m hours at v km/h. It then travels n hours at u km/h. How far does it travel in all?

18   What is the total number of pence in

£a + b pence?

19   A train travels at v km/h. How far does it go in x hours and how long does it take to go y km?

20   Two numbers, less than 10, are chosen. Add them; multiply this sum by 2; add 4; multiply by 3; add 4 times one of the original numbers; take away 12; take away 5 times the second of the original numbers. The result is 10m + n. What were the original numbers? Interpret this result as a party trick.

2

Elementary Operations in Algebra

10 Use of symbols

In order that you may become familiar with the processes of algebra you need considerable practice in the use of symbols. Consequently in this and subsequent chapters you will constantly be using letters which represent numbers in a general way and without any reference to quantities such as length, cost, etc., as was done in the previous chapter.

Thus when we employ the form a + b we shall, in general, be using the letters, not as referring to any particular quantity, but as standing for any numbers.

11 Symbols of operation

As stated in Section 1, certain symbols of operation, such as +, −, ×, ÷, $\sqrt{}$, are common to arithmetic and algebra, since they are used for operations which are performed in both subjects. Usually, however, there is a certain difference in the way they are employed. It is evident that while such operations as

$$5 + 7, 10 - 3, 6 \times 4, 15 \div 3, \sqrt{9}$$

can be, and usually are carried out at once with definite numerical results, expressions such as

$$a + b, a - b, a \times b, a \div b, \sqrt{a}$$

cannot be evaluated numerically while a and b represent any numbers. Until numerical values are assigned to them we cannot proceed further with the operation. But we can, and do, operate with the expressions themselves, without any reference to their numerical values.

In addition to the above, many other symbols of operation are used in Algebra, among them the following:

Symbol	Meaning
=	See Section 6
≠	is not equal to
≃	is approximately equal to
>	is greater than
<	is less than

12 Algebraic expression. Terms

Such a combination of letters and symbols as 2a + b is an example of what is called an algebraic expression. It may be defined as follows:

An algebraic expression is a combination of symbols which stand for numbers and for operations with them.

For brevity the term 'expression' is usually employed. When the expression contains the symbols of operation + or −, those parts of the expression which they separate are called terms.

Thus 2a + 3b is an expression of two terms or a binomial.

$\frac{5x}{2} - \frac{3y}{5} + 6z$ is an expression of three terms or a trinomial.

A combination of letters which does not contain either of the signs + or − may be said to be an expression of one term, or a monomial.

Thus $\frac{5ab}{6}$ is such an expression.

13 Brackets

It frequently happens that an expression or part of an expression, is to be operated as a whole. For example, suppose we wish to write in algebraic symbols 'Twice the sum of a and b'.

Evidently the arrangement 2 × a + b does not make it clear whether the 2 is to multiply a only or the sum of a and b.

Consequently we employ 'brackets' to enclose the part which is to be operated on as a whole – viz., a + b.

∴ we write 2(a + b).

In this arrangement the multiplication sign is omitted between the 2 and the bracket.

The brackets have the effect of indicating the order in which operations are to be carried out. Thus:

$$2(a + b) - c$$

means that we find the sum of a and b, multiply this by 2 and then subtract c.

Similarly (a + b) × (c + d) or (a + b)(c + d) means that we find the sum of a and b, and also of c and d and then multiply the two results.

This will be considered further at a later stage.

14   Coefficient

The expression 3a denotes a multiple of a and the number 3, which indicates the multiple, is called the coefficient of a.

The coefficient may be a definite number like 3, called a numerical coefficient, or it may be a letter representing a number.

Thus in the expression ax, a may be regarded as the coefficient of x, but in some problems when we are thinking of multiples of a, x would be a coefficient of a; in such a case we would usually write the expression as xa.

In general, if an expression is the product of a number of factors, any one of them can be regarded as the coefficient of the product of the others, when for any purpose we regard this product as a separate number.

Thus in 3ab, 3 is the coefficient of ab
    3a is the coefficient of b
    3b is the coefficient of a.

In an expression, terms which involve the same letter, and differ only in the coefficients of this letter, are called **like terms**.

Thus in the expression

3a + 5b − 2a + 4b
3a and 2a are like terms

and   5b and 4b are like terms.

15 Addition and subtraction of like terms

In arithmetic we learn that the sum of

	5 dozen and 9 dozen is 14 dozen,	
or	$(5 \times 12) + (9 \times 12) = 14 \times 12.$	(A)

Similarly	8 score $+$ 7 score $=$ 15 score
or	$(8 \times 20) + (7 \times 20) = 15 \times 20.$

So for any number as for example

$$(6) \times 24 + (11 \times 24) = 17 \times 24.$$

In algebra, if we were to let a represent 12 in the statement (A) given at the beginning of this section, we could write

$$5a + 9a = 14a,$$

and for the other cases

	$8a + 7a = 15a$
and	$6a + 11a = 17a.$

These last three cases are generalised forms of the preceding examples, but it must be noted that whereas in the arithmetical forms we can proceed to calculate the actual value of the sum in each, in the algebraical forms we can proceed no further in the evaluation until a definite numerical value is assigned to a.

Subtraction leads to similar results, just as

	9 dozen $-$ 5 dozen $=$ 4 dozen.
so	$9a - 5a = 4a.$

In this way we can add or subtract like terms only. It is not possible, for example, to perform any addition of two unlike terms such as $9a + 5b$.

The rule for adding together like terms will now be clear. It is 'add the coefficients'.

Thus the sum of $2x + 5x + 3x = 10x$, whatever x may be. The operation of 'finding the sum' is used to include both addition and subtraction. This is called the 'algebraic sum'.

When an expression contains more than one set of like terms, these should be collected and dealt with separately.

16 Worked examples

Example 1 Simplify $5a + 6b + 2a - 3b$.

Collecting like terms $5a + 2a = 7a$
$$6b - 3b = 3b.$$

Hence the whole expression is equal to $7a + 3b$. In practice there is generally no need to write down the above steps. The calculations can be made mentally.

Example 2 Simplify $15x - 3y + 6y + 7x - 5$.

Collecting like terms and adding coefficients we get

$$22x + 3y - 5.$$

17   The order of addition

The counting of a number of things is not affected by the order in which they are counted. Thus:

$$6 \text{ apples} + 4 \text{ apples}$$

is the same in number as

$$4 \text{ apples} + 6 \text{ apples}.$$

This will be clear when it is remembered that 6 is the symbol for 6 units, and 4 is the symbol for 4 units.
Thus:

$$6 + 4 = (1 + 1 + 1 + 1 + 1 + 1) + (1 + 1 + 1 + 1)$$

and

$$4 + 6 = (1 + 1 + 1 + 1) + (1 + 1 + 1 + 1 + 1 + 1).$$

In each case the total number of units is the same.

Thus algebraically $6a + 4a$ is the same in value as $4a + 6a$. This is true for any algebraical sum.

Thus $6a + 5b - 3a$ can be written as $5b - 3a + 6a$ without altering the value of the expression.

Briefly the order in which numbers may be added is immaterial.

18   Evaluation by substitution

If we wish to find the numerical value of an algebraical expression for definite numerical values of the letters composing it, the expression should first be simplified by adding like terms. Then the numerical values are substituted for the letters.

Example: Find the value of $6x + 2y - 3x + 4y - 3$ when $x = 3$ and $y = 2$.

Simplify the expression as in Section 14:

$$6x + 2y - 3x + 4y - 3 = 3x + 6y - 3.$$

Substituting the given values:

$$\begin{aligned}
3x + 6y - 3 &= (3 \times 3) + (6 \times 2) - 3 \\
&= 9 + 12 - 3 \\
&= 18.
\end{aligned}$$

Note: It will be seen that brackets are introduced when it is desirable to keep terms separate for evaluation.

Exercise 2

1 (a) Find the value of 6 dozen + 4 dozen.
  (b) Simplify $6a + 4a$ and find its value when $a = 12$.
2 (a) Find the value of $(8 \times 73) - (3 \times 73)$.
  (b) Find the simplest form of $8b - 3b$ and find its value when $b = 73$.
3 Write down in its simplest form:

$$a + a + a + b + b + b + b$$

and find its value when $a = 5$ and $b = 8$.
4 Add together $2a$, $4a$, a, $5a$ and $7a$ and find the value of the sum when $a = 2.5$.
5 Write the following expressions in their simplest forms:

  (1) $15b + 11b$.   (2) $15x - 3x + 7x$.
  (3) $9a - 4a + 6a + a$.   (4) $4x + 3x - 2x - x$.

6 Write in their simplest forms:

  (1) $5a - 2b - 3a + 6b$.   (2) $11p + 5q - 2q + p$.
  (3) $a - 2 + 3b + 6 + 5a$.

7 Add together:

  (1) $4a - 5b$, $a + 6b$, $5a + b$.
  (2) $b + c - 3d$, $c + 2b + d$, $d - b - c$.
  (3) $5x + 2y + 3z$, $x - y - 2z$, $2x - y + z$.

8 When $a = 2$, $b = 3$ find the numerical values of

  (1) $3a + 2b + 1$.   (2) $5a - 3b + 6$.
  (3) $6a + 2b - 3a + 1$.   (4) $4a - 5b - 2b + 12a$.

9   Simplify the following expressions and find their values when
a = 4, b = 2, x = 3, y = 5.

(1) 4ab − 2ab + 6ab.
(2) 5ax − 2ax + bx.
(3) 6xy − 4xy + xy.
(4) ab + 6bx − ay + 3ab − 2bx.

10   Find the numerical value of $a + \frac{1}{2}a + \frac{1}{4}a + \frac{1}{8}a$ when a = 2.

11   When x = 1, y = 2, z = 3, find the numerical value of

$$3x + 5y − 4z + 8x − y + 5z.$$

12   When x = 4, y = 5, z = 1, find the values of

(1) 3xy + 2yz − 3z − 1.   (2) xy + yz + zx.
(3) xy + y + x + 1.

13   If n be an odd number, write down the next three odd
numbers greater than it and find their sum.

14   Write down a series of four numbers of which the first is a,
and each of the others is twice the one which precedes it. Find
their sum.

15   Write down a series of 5 numbers of which the first is a, the
second is greater than the first by d and each of the other
three is greater by d than the one which precedes it. Find their
sum.

16   There are 5 numbers, the smallest of which is expressed by
2n + 5. Each of the others is 3 greater than the one which
precedes it. Write down the numbers and find their sum.

19   Multiplication

Order of multiplication.   In algebra, as in arithmetic, the multiplication of a number of factors may be performed in any order,
or, more precisely:

*The product of a number of factors is independent of the order
in which they are multiplied*

Thus   3 × 4 is equal in value to 4 × 3
          6 × 3 × 5 is equal in value to 3 × 5 × 6

and generally   a × b is equal in value to b × a
and   a × b × c is equal in value to c × b × a.

Consequently if it is required to multiply say, 2a by 5 we can write the product thus:

$$2 \times a \times 5 = 2 \times 5 \times a$$
$$= 10a$$

and
$$3a \times 2b = 3 \times 2 \times a \times b$$
$$= 6ab.$$

In this last example, a and b being unlike letters, we cannot proceed further with the multiplication.

It should be noted, however, that although 4×3 is equal in value to 3×4, the two products do not necessarily mean the same thing when they refer to quantities.

If, for example, 12 soldiers were to 'form fours', they would be arranged as shown in Fig. 1(a), the arrow showing the direction in which they are facing. But if the same 12 men were to 'form threes' they would be arranged as in Fig. 1(b).

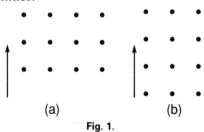

(a)   (b)

Fig. 1.

Thus 3 rows of 4 men require the same number of men as 4 rows of 3 men, but they are a different arrangement. Similarly if 4 men pay 3 pence each; the total amount paid is the same as that when 3 men pay 4 pence each.

20 Powers of numbers

The product of equal numbers is called a **power**. Thus:

8×8 is called the second power of 8, or the square of 8.

$8 \times 8 \times 8$ is called the third power of 8, or the cube of 8, there being three equal factors.

$8 \times 8 \times 8 \times 8$ is called the fourth power of 8, there being four equal factors.

Similarly:

$a \times a$ is the second power of a or the square of a.

$a \times a \times a$ is the third   power of a or the cube of a.

$a \times a \times a \times a$ is the fourth power of a.

The process of writing a power in full is tedious, and the form of it restricts further operations, especially when the power is a high one. Accordingly mathematicians made many attempts through centuries to devise a symbolic method of representing the row of factors. Finally, Descartes in 1637 used a numeral to mark the number of factors or the power and wrote the cube of a for example as a^3, the fourth power as a^4, etc.

The figure used in this way is called an **index** or **exponent**; it indicates the number of factors.

Thus $a \times a \times a \times a \times a$ is written as a^5. With this symbolic method it is as easy to write down the 20th power of a number as the 2nd.

But a new symbol, if it is to be satisfactory, must not only express clearly and concisely the purpose for which it was devised, but it must also be convenient for operations with it. We shall see, from what follows, that an index fulfils this condition, and later in the book it will be seen how it lends itself to important developments.

21 Multiplication of powers of a number

Suppose we require to multiply two powers of a – say $a^2 \times a^3$. These numbers written in full are

$$(a \times a) \times (a \times a \times a)$$

the brackets serving to separate the two powers.

It is a fundamental law of algebra, which will be assumed here, that when two groups of factors are multiplied, the factors in the groups are associated as one group of factors to give the product.

By this law
$$(a \times a) \times (a \times a \times a)$$
$$= (a \times a \times a \times a \times a).$$

∴ the number of factors in the product is the sum of the number of factors in the two groups.

∴ in the example above the product $(a \times a \times a \times a \times a)$ is the 5th power of a, and the index of the product is the sum of the indices of the two factors.

∴
$$a^2 \times a^3 = a^{2+3}$$
$$= a^5.$$

The same reasoning may be applied to other cases and so we may deduce the general rule for the multiplication of two powers of a number.

When two powers of the same number are multiplied, the index of the product is the sum of the indices of the factors.

Examples

(1) $x^4 \times x^4 = x^{4+4} = x^8$.
(2) $2a^7 \times a^3 = 2a^{7+3} = 2a^{10}$.
(3) $5b^2 \times 3b^5 = 5 \times 3 \times b^2 \times b^5 = 15 \times b^{2+5} = 15b^7$.
(4) $a^b \times ab^2 = a^2 \times a \times b \times b^2 = a^3b^3$.

The rule may be extended to the product of more than two factors. Thus:

$$x^2 \times x^2 \times x^4 = x^{2+3+4}$$
$$= x^9.$$

22   Power of a product

To find the value of $(ab)^2$.

The use of the bracket shows that, as stated in Section 11, the expression within the bracket must be regarded as a whole.

∴ by definition of a power

$$(ab)^2 = (ab) \times (ab)$$
$$= (a \times a) \times (b \times b)$$
$$= a^2b^2.$$

Thus we see that the effect of this is that the index 2 must be distributed over each of the factors.

Thus: $$(2xy)^3 = 2^3 \times x^3 \times y^3$$
$$= 8x^3y^3.$$

So in arithmetic $(2 \times 5)^2 = 2^2 \times 5^2 = 4 \times 25 = 100$. Consequently when taking a power of a product, the index of the power is said to be distributed and applied to each factor of the product.

Exercise 3

Write down the following in their simplest forms:

1 $4a \times 3$ **2** $5x \times 2y$
3 $\frac{1}{2}x \times 4y$ **4** $7m \times 3n$
5 $\frac{1}{3}a \times \frac{1}{4}b$ **6** $6a \times \frac{2}{3}b$

7 $3a \times 4b \times 5c$ **8** $\frac{x}{2} \times \frac{y}{3} \times \frac{z}{4}$

9 $x^2 \times x$ **10** $a \times a^2 \times a$
11 $x^2 \times x^2$ **12** $a^3 \times a^3$

13	$2a^2 \times a^3$	**14**	$3x^3 \times 2x^4$
15	$2ab \times ab$	**16**	$2b \times 3b^8$
17	$x^2y \times xy^2$	**18**	$7x^3a \times x^3a$
19	$2a \times 3a^2 \times a^3$	**20**	$(3a^2b)^3$
21	$(x^3)^2$	**22**	$(2a^4)^3$
23	$(2a^3)^4$	**24**	$(4a)^2 \times 4a^2$

Find the numerical values of the following:

25 $2a^2 \times a$, when $a = 3$
26 $2a^2 + a$, when $a = 3$
27 $a^2b \times ab^2$, when $a = 1$, $b = 2$
28 $x^2 + 7x + 2$, when $x = 10$
29 $3a^2 + 2ab + b^2$, when $a = 2$, $b = 3$
30 $c^2 \times c^3$, when $c = 1$
31 $3a \times 3a \times 3a$, when $a = 2$
32 $(2a^2x)^2$, when $a = 2$, $x = 3$

---

23 Division of powers

Suppose that a power of a number is divided by another power of the same number, as for example,

$$a^5 \div a^2.$$

Every division can be expressed in fractional form, as in arithmetic.

$$\therefore \qquad a^5 \div a^2 = \frac{a^5}{a^2}$$
$$= \frac{a \times a \times a \times a \times a}{a \times a}$$

(by definition of a power).

As in arithmetic, common factors in the numerator and denominator can be cancelled.

∴ the two a factors in the denominator can be cancelled with two of the five factors of the numerator.

Then there will be left in the numerator $(5 - 2)$ factors.

$$\therefore \qquad a^5 \div a^2 = a^{5-2}$$
$$= a^3.$$

Clearly the same method can be followed whatever the powers. Consequently we may deduce the rule:

When dividing a power of a number by another power of the same number, subtract the index of the divisor from the index of the dividend.

Example: Divide $84a^6$ by $12a^2$.

$$\frac{84a^6}{12a^2} = 7a^{6-2} = 7a^4.$$

Example: Divide $3x^4$ by $6x^6$.

Arranging as above.

$$3x^4 \div 6x^6 = \frac{3x^4}{6x^6}$$

$$= \frac{3 \times x \times x \times x \times x \times x \times x}{6 \times x \times x \times x \times x \times x \times x \times x \times x \times x}.$$

In this example as the higher power is in the denominator, on cancelling there are $(6 - 4)$ factors in the denominator. Hence we get:

$$\frac{1}{2x^{6-4}} = \frac{1}{2x^2}.$$

Exercise 4

Write down answers to the following:

1 $a^3 \div a$	**2** $a^4 \div a^2$
3 $3x^3 \div x^2$	**4** $b^5 \div 2b$
5 $6a^6 \div 3a^2$	**6** $5y^4 \div y$
7 $6x^7 \div 2x^4$	**8** $14c^6 \div 7c^4$
9 $a^2b^3 \div ab^2$	**10** $x^5y^4 \div x^2y^3$
11 $5a^3b^3 \div ab^2$	**12** $6x^4y \div 2x^3$
13 $6a^4 \div 3a^4$	**14** $15x^3 \div 12x^3$

24   Easy fractions

Algebraic fractions obey the same fundamental laws as fractions in arithmetic. In principle they are manipulated by the same methods. But since the numerators and denominators may be algebraical expressions, sometimes rather complicated, they present difficulties not found in arithmetic fractions. In this chapter we shall deal only with simple forms involving easy manipulation, more difficult cases being left until later.

25 Addition and subtraction

The methods of arithmetic can readily be applied, as shown in the following examples.

Example 1: Find the sum of $\frac{x}{3} + \frac{x}{5}$.

This is of the same form as

$$\frac{2}{3} + \frac{2}{5}$$

and is worked in the same way.

Just as
$$\frac{2}{3} + \frac{2}{5} = \frac{(2 \times 5) + (2 \times 3)}{3 \times 5}$$

so
$$\frac{x}{3} + \frac{x}{5} = \frac{(x \times 5) + (x \times 3)}{3 \times 5}$$
$$= \frac{5x + 3x}{15} = \frac{8x}{15}.$$

Example 2: Find the sum of $\frac{3}{a} + \frac{4}{b}$.

This is similar in type to the preceding and is dealt with in the same way:

$$\frac{3}{a} + \frac{4}{b} = \frac{(3 \times b) + (4 \times a)}{ab}$$
$$= \frac{3b + 4a}{ab}.$$

Example 3: Simplify $\frac{x}{y} - \frac{a}{b}$.

Proceeding as before

$$\frac{x}{y} - \frac{a}{b} = \frac{(x \times b) - (a \times y)}{y \times b}$$
$$= \frac{bx - ay}{by}.$$

Example 4: Simplify $\frac{2a}{15} + \frac{5b}{12}$.

As in arithmetic, we find the L.C.M. of the denominators – viz. 60.

Then
$$\frac{2a}{15} + \frac{5b}{12} = \frac{(2a \times 4) + (5b \times 5)}{60}$$
$$= \frac{8a + 25b}{60} .$$

Note: It is not possible to cancel any factors of 8 or 25 with factors of 60. This mistake is sometimes made by beginners, although the proceeding is contrary to the laws of arithmetic. Only factors common to each term of the numerator can be cancelled with factors of the denominator.

Example 5: Simplify $\frac{x}{12a^2b} - \frac{y}{18ab^2}$.

We must find the L.C.M. of the denominators. To do this we find the L.C.M. of the numerical coefficients 12 and 18 – i.e., 36 – then the L.C.M. of a^2b and ab^2. This is a^2b^2, since both of them will divide into it exactly. The product of 36 and a^2b^2 is the L.C.M. of the denominators.

$$\therefore \qquad \frac{x}{12a^2b} - \frac{y}{18ab^2} = \frac{(x \times 3b) - (y \times 2a)}{36a^2b^2}$$
$$= \frac{3bx - 2ay}{36a^2b^2} .$$

26　Multiplication and division

As in arithmetic, these operations are based upon the same important rule of fractions, viz.:

If numerator and denominator are divided by the same number, the value of the fraction is unaltered.

This can be expressed algebraically as follows:

Let $\frac{a}{b}$ be any fraction.

Then
$$\frac{a}{b} = \frac{a \times n}{b \times n}$$

and
$$\frac{a}{b} = \frac{a \div m}{b \div m} ,$$

where m and n are any numbers.

Example 1: Simplify $\frac{4x^3y}{6xy^3}$.

Written in full the fraction is:

$$\frac{4x^3y}{6xy^3} = \frac{4 \times x \times x \times x \times x \times y}{6 \times x \times y \times y \times y}.$$

Cancelling common factors this is equal to:

$$\frac{2 \times x \times x}{3 \times y \times y} = \frac{2x^2}{3y^2}.$$

Note: In practice there is no need to write out the powers in full as shown above. The rule for the division of powers may be applied directly.

Example 2: Simplify $\frac{6ax^4}{14x^2y^2} \times \frac{2y^3}{3a^4}$.

As in arithmetic, factors in either numerator can be cancelled with factors in either denominator.

$$\therefore \qquad \frac{6ax^4}{14x^2y^2} \times \frac{2y^3}{3a^4}.$$
$$= \frac{x^{4-2}}{7} \times \frac{2y^{3-2}}{a^{4-1}}.$$
$$= \frac{x^2 \times 2y}{7 \times a^3}$$
$$= \frac{2x^2y}{7a^3}.$$

Example 3: Simplify $\frac{8x^3}{5a^2y} \div \frac{4x^2}{3a}$.

Proceeding as with a similar arithmetical example:

$$\frac{8x^3}{5a^2y} \div \frac{4x^2}{3a} = \frac{8x^3}{5a^2y} \times \frac{3a}{4x^2}$$
$$= \frac{2x \times 3}{5ay}$$
$$= \frac{6x}{5ay}.$$

Exercise 5

Simplify the following:

1 $\frac{a}{5} + \frac{a}{7}$

2 $\frac{2x}{3} - \frac{3x}{5}$

3 $\dfrac{4}{x} + \dfrac{5}{y}$

4 $\dfrac{3a}{2b} + \dfrac{5a}{3b}$

5 $\dfrac{9x}{10y} - \dfrac{5x}{8y}$

6 $\dfrac{7x}{3y} + \dfrac{2x}{9y} - \dfrac{5x}{6y}$

7 $\dfrac{5a}{6bc^2} + \dfrac{3b}{8a^2c}$

8 $x + \dfrac{4}{3x}$

9 $3ab + \dfrac{5}{2b}$

10 $8 - \dfrac{6}{5y^2}$

11 $a + 1 + \dfrac{1}{a}$

12 $x - 1 + \dfrac{1}{x}$

13 $\dfrac{1}{ab} + \dfrac{2}{bc} + \dfrac{3}{ac}$

14 $\dfrac{5}{a^2bc} - \dfrac{4}{ab^2c} + \dfrac{2}{abc^2}$

---

15 $\dfrac{x^4y^2}{x^2y}$

16 $\dfrac{12a^2b}{3a^4}$

17 $\dfrac{15a^2b^4c}{10a^4b^2c^2}$

18 $\dfrac{x^3 \times x^3}{x^6}$

19 $\dfrac{4x^4 \times 2x^3y}{8x^6}$

20 $\dfrac{a^4 \times bc^2}{a^3 \times b^2c^3}$

21 $\dfrac{2a}{9} \times \dfrac{6}{a^2}$

22 $\dfrac{x^2}{xy} \times \dfrac{3y^2}{2x}$

23 $\dfrac{3p}{2q^2} \times \dfrac{6q}{5p}$

24 $2x^2 \times \dfrac{y}{3x}$

25 $2x \times \dfrac{x}{y}$

26 $2x \div \dfrac{x}{y}$

27 $\dfrac{x}{y} \div \dfrac{x}{y}$

28 $\dfrac{x}{y} \div \dfrac{y}{x}$

29 $\dfrac{3a}{7} \div \dfrac{9a}{14}$

30 $\dfrac{2x^2}{y} \div \dfrac{3y}{x^2}$

31 $\dfrac{6x^2}{8yz} \div \dfrac{3xy}{4xz}$

32 $\dfrac{a^5}{b^4} \div \dfrac{3a^2}{2b^2}$

33 $\dfrac{4abc}{5b^2} \div \dfrac{3c^2}{10ab}$

34 $3xyz = \dfrac{4xz}{y}$

35 $30 \div \dfrac{3xyz}{2z}$

36 $\dfrac{ab}{c^2} \div \dfrac{c^2}{4ab}$

3

Brackets and Operations with Them

27 Removal of brackets

Simple examples of the use of brackets as a convenient way of grouping numbers have already been considered. In this chapter we will examine extensions of their use, operations with them and the simplification of algebraical expression, which contain brackets, by removing them.

We will begin with a simple but important case.

Fig. 2 represents a rectangle made up of two other rectangles.

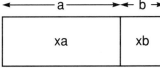

Fig. 2.

Let a mm = the length of one rectangle

Let b mm = the length of the other rectangle

Let x mm = the breadth of each rectangle as shown in Fig. 2.

Then $(a + b)$ mm = the length of the combined rectangle, placing $(a + b)$ in brackets shown in Section 13.

The areas of two smaller rectangles are xa and xb mm^2 and area of whole rectangle = $x(a + b)$ mm^2.

But the area of the whole rectangle equals the sum of the area of the parts.

∴ $$x(a + b) = xa + xb.$$

Similarly if there are three rectangles of lengths a, b and c ins respectively, then in the same way we could show that:

$$x(a + b + c) = xa + xb + xc.$$

By modifications of the figures, which are left to the ingenuity of the student, we could similarly show that:

(1)　　$x(a - b) = xa - xb$.
(2)　$x(a + b - c) = xa + xb - xc$.
(3)　$x(a - b + c) = xa - xb + xc$.

In all these examples expressions containing brackets have been transformed into expressions without brackets, or, as we say, the brackets have been removed.

Hence we can deduce that:

When the whole of an expression within brackets is multiplied by any number, then, if the brackets be removed, each term within the brackets must be multiplied by the number.

The factor without the brackets is said to be distributed as a factor of each term within the brackets.

This is an example of the algebraical law called the Law of distribution.

28　Addition and subtraction of expressions within brackets

There are four cases which we will consider. They are represented by the following expressions:

(1) $a + (b + c)$.
(2) $a + (b - c)$.
(3) $a - (b + c)$.
(4) $a - (b - c)$.

The question to be considered is, What will be the effect of removing the brackets in the above expressions, a, b and c representing any numbers?

(1) $a + (b + c)$

Let a, b, c be represented by the areas of rectangles as shown in Fig. 3.

Fig. 3.

It is evident that the area of the whole rectangle which represents a + (b + c) is the sum of the three rectangles representing a, b, c.

∴ $\qquad$ a + (b + c) = a + b + c

The steps in the addition of the numbers in this case are not altered by the insertion or the removal of the brackets.

(2) a + (b − c)

As in the previous case, a, b, c are represented by the areas of rectangles as shown in Fig. 4.

Fig. 4.

The two unshaded rectangles represent a and b − c.

∴ the whole unshaded portion represents a + (b − c). It can be formed either:

(1) by adding (b − c) to a.
(2) by adding b to a and then subtracting c.

These two results are equal.

∴ $\qquad$ a + (b − c) = a + b − c

Thus no change results when the brackets are removed.

(3) a − (b + c)

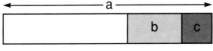
Fig. 5.

Using the same method as before of representing a, b and c, the unshaded rectangle (Fig. 5) represents the remainder when (b + c), the two shaded rectangles, are subtracted from a, the whole rectangle, i.e., it represents a + (b + c).

Also if from a, the whole rectangle, we subtract b and c in turn the remaining rectangle is the unshaded portion.

∴ it represents a − b − c.

∴ $\qquad$ a − (b + c) = a − b − c.

(4) a − (b − c)

In Fig. 6 the rectangle representing a, b, c is shown, a being represented by the whole rectangle.

The shaded rectangle represents (b − c).

The unshaded rectangle represents the result of subtracting (b − c) from a, i.e., it represents a − (b − c).

Fig. 6.

It may also be considered as representing the result of subtracting b from a and then adding c, i.e. it represents a − b + c.

∴ $$a - (b - c) = a - b + c.$$

Collecting the four cases we have:

(1) a + (b + c) = a + b + c.
(2) a + (b − c) = a + b − c.
(3) a − (b + c) = a − b − c.
(4) a − (b − c) = a − b + c.

From these results we can deduce two rules respecting signs when the brackets are removed.

A. *From (1) and (2) when the + sign precedes the brackets the signs of the terms within the brackets are unaltered.*

B. *From (3) and (4) when the − sign precedes the brackets the signs of the terms within the brackets are changed.*

29 Worked examples

The following are examples of the use made of the rules of Section 27 and Section 28 when brackets are removed from an algebraic expression in order to simplify it.

Example 1: Simplify $a(a^2 + ab + b^2)$.

When the brackets are removed the rule of Section 27 is employed and the factor a multiplies each term within the brackets.

Thus: $a(a^2 + ab + b^2) = a^3 + a^2b + ab^2.$

Example 2: Simplify $2(4a + 3b) + 6(2a - b)$.

When removing the brackets we use Section 27 to multiply by 2 and 6, and from Section 28, since the + sign before the second pair of brackets is positive, there is no change of sign.

$$2(4a + 3b) + 6(2a - b)$$
$$= 8a + 6b + 12a - 6b$$
$$= 20a.$$

Example 3: Simplify $5x - (5y + 2x)$.

This is an example of case (3) of Section 28. On removing the brackets signs are changed.

Thus
$$5x - (5y + 2x)$$
$$= 5x - 5y - 2x$$
$$= 3x - 5y \text{ (on adding like terms).}$$

Example 4: Simplify $3(4a - b) - 2(3a - 2b)$.

This involves the rule of Section 27 and case (4) of Section 28. Using these

$$3(4a - b) - 2(3a - 2b)$$
$$= 12a - 3b - 6a + 4b$$
$$= 6a + b$$

Example 5: Simplify $x(2x - y) - x(x - y) - y(x + 2y)$ and find its value when $x = 2$, $y = 1$.

$$x(2x - y) - x(x - y) - y(x + 2y)$$
$$= 2x^2 - xy - x^2 + xy - xy - 2y^2$$
$$= x^2 - xy - 2y^2 \quad (\text{since } + xy - xy = 0).$$

Substituting $x = 2$, $y = 1$.

$$x^2 - xy - 2y^2 = (2)^2 - (2 \times 1) - 2(1)^2$$
$$= 4 - 2 - 2$$
$$= 0.$$

Note: This example shows the advantage of simplifying the expression before substituting the values of x and y.

Exercise 6

Simplify the following expressions by removing brackets.

1 $3(5x + 6z)$
2 $2a(3a + 4b)$
3 $6a^2(3a + 7b - 6c)$
4 $2(x + 2y) + 3(2x - y)$
5 $x(x^2 - 3x) + x^2(4x + 7)$
6 $\frac{1}{2}(x - 2y) + \frac{1}{4}(2x + 4y)$
7 $2(x + y + z) + 3(2x + y - 2z)$
8 $x - (2y + z)$
9 $2x - (y - 2z)$
10 $2(2a + 2b) - 3(a - b)$
11 $3a - (2a + b)$

12 $3a - (2a - b)$
13 $5x - (x - 2y + 2z)$
14 $3(a + b - c) - 2(a - b + c)$
15 $4(x + y) - 3(2x - y) + 2(x - 2y)$
16 $a(a + b) - b(a - b)$
17 $x^2(x + y) - xy(x^2 - y^2)$
18 $3(x^2 + x + 5) - 2(x^2 - 3x - 4)$
19 $2p(3p + 2q) - 3q(2p - 5q) + p(3p + 5q)$
20 $5(xy)^2 - 3x(y - 2x)$
21 $(2x^2)^2 - 2x^3(x - 4)$

30 Systems of brackets

It may happen that an expression within brackets is part of another expression which is itself within brackets. In that case a second set of brackets would be required, and to avoid confusion they must be of a shape different from those already used, such as $\{-\}$ or $[-]$.

For example:

$$40 - \{2(a + b) + 5(a - b)\}.$$

You will easily recognise how clearly and effectively the brackets help to show the construction of the expression and relations of the different parts to one another.

It might happen that the whole of the above expression is to be multiplied by $2b$. This will necessitate another set of brackets which will indicate that the expression is to be treated as a whole. We would express this as follows:

$$2b[40 - \{2(a + b) + 5(a - b)\}].$$

Sometimes a straight line may be placed over a part of an expression with the same meaning as brackets.

Thus $x - \overline{y + z}$ has the same meaning as $x - (y + z)$. It must also be remembered that an expression which is the numerator of a fraction must be regarded as a whole and must be treated as if it were in a bracket.

Thus $\qquad\qquad 5a - \dfrac{a - b}{2}$

means the same as $\qquad 5a - \tfrac{1}{2}(a - b)$.

If the whole expression were multiplied by 2, it would become $10a - (a - b)$.

When expressions with two or more sets of brackets are to be simplified by removal of the brackets, it is best, as a rule, to begin with the innermost and work outwards. Examples of this will be seen in the following.

31 Worked examples

Example 1: Simplify $2\{3a + 5(b + c)\}$.

As stated above we begin by removing the inner brackets.

$$2\{3a + 5(b + c)\}$$
$$= 2\{3a + 5b + 5c\}$$
$$= 6a + 10b + 10c.$$

Example 2: Simplify $3\{3a - 2(a - b)\}$.

$$3\{3a - 2(a - b)\}$$
$$= 3\{3a - 2a + 2b\}$$
$$= 3\{a + 2b\}$$
$$= 3a + 6b.$$

Example 3: Simplify $12a - 2[3a - \{4 - 2(a - 3)\}]$.

Beginning with the innermost bracket

$$12a - 2[3a - \{4 - 2(a - 3)\}]$$
$$= 12a - 2[3a - \{4 - 2a + 6\}]$$
$$= 12a - 2[3a - \{10 - 2a\}] \quad \text{(adding like terms)}$$
$$= 12a - 2[3a - 10 + 2a]$$
$$= 12a - 2[5a - 10]$$
$$= 12a - 10a + 20$$
$$= 2a + 20.$$

Exercise 7

Remove the brackets from the following expressions and simplify them.

1 $3\{5a - 3(a + 1)\}$
2 $3\{4(a + b) - 3(a - 2b)\}$
3 $\frac{1}{2}\{6x - 3(2x - 1)\}$
4 $5a^2 + 2a\{b - (a + c)\}$
5 $3p^2 - \{2p^2 - p(p + 1)\}$

6 $3x(x + 3y) - 2\{x^2 + 3y(x - 2y)\}$

7 $3bc - 2\{b(b - c) - c(b + c)\}$

8 $15x - [3x - \{2x - (x - 5)\}]$

9 $50 - 2[3a + 2\{3b - 4(b - 1)]\}$

10 $2(x + y) - \overline{x - y}$

11 $12\left\{\dfrac{a - b}{2} - \dfrac{a + b}{3}\right\}$

12 $3x\left\{\dfrac{2x - y}{3} - \dfrac{x - 2y}{6}\right\}$

13 $3c - \left\{\dfrac{2a + c}{8} - \dfrac{a + 2c}{4}\right\}$

Fill in the blanks within the brackets in the following:

14 $2a - b + c = 2a - ($ $)$

15 $x - y - z = x - ($ $)$

16 $2a + 4b - 6c = 2($ $)$

17 $x^2 - xy + y^2 = x^2 - y($ $)$

18 From $3a - 2b + 4c$ subtract $a + 2b - 3c$

19 Take $2x - 3y + 4z$ from $3x - y + 2z$

20 When $a = 3$, $b = 2$, $c = 1$, find the values of the following:
    (1) $4a(a + 4b) - a(3a - b)$
    (2) $3c\{4c - (3c - 1)\}$

4

Positive and Negative Numbers

32   The scale of a thermometer

Figs. 7a and 7b represent portions of Celsius thermometers in which a fine column of mercury registers the rise and fall of temperature.

The zero point, marked 0, indicates the position of the mercury in the tube at freezing point – i.e., the freezing point of water.

Fig. 7(a) shows the mercury at 8°C above zero.

Now suppose the temperature falls 16C° below this point.

First it falls 8C° to 0°C, and then continues to fall for 8C° below zero. To show this temperature on the scale it must be marked in some way which is different from the 8°C above zero, or there would be confusion. To distinguish the degrees below zero from those above we use the plan of putting a minus sign, −, before all those below zero, and if necessary a plus sign, +, before those above zero.

Thus +8°C means 8 degrees above zero, and −8°C means 8 degrees below zero.

We may call these positive and negative degrees, using the terms, and the signs + and −, to indicate different directions up and down from the zero.

(a)   (b)

Fig. 7.

33 An example from time

In reckoning time, the numbers denoting the years are counted from the birth of Christ. Years after that event are denoted by AD, as AD 1941, and those before by BC, as 55 BC. The use of the symbols AD and BC is, in principle, similar to the use of + and − in the case of the thermometer.

34 A commercial illustration

In a certain transaction at market a farmer made a profit of £12.

In a second deal he lost £8.

Consequently in the two transactions he made a net gain of £12 − £8 = £4.

In a third transaction he lost £10.

His profit and loss account is now shown by £4 − £10.

If this loss had been £4, his position would be £4 − 4 = 0 – i.e., he has reached a zero position, neither loss nor gain.

But he lost £10, not £4; therefore he has a net loss of £6 – i.e., he is £6 below his zero.

To distinguish gains from losses, we could, as in the case of the thermometer, place the

negative sign before amounts showing losses,

and the

positive sign before amounts showing profits.

In that sense £4 − £10 = − £6, the negative sign indicating a loss of £6.

35 Motion in opposite directions

Suppose a man starting from a point O (see Fig. 8) travels for 4 km in the direction O to X, reaching the point marked A.

Fig. 8.

He then turns and travels 6 km in the opposite direction, O to X¹. After 4 km he reaches O, his zero or starting point. The next 2 km take him to B.

He is now 2 km from O but in the direction opposite to that in which he started. His succesive distances from O can be shown by $+4 - 6$. This suggests that, as in the previous cases, if distances from O in one direction were regarded as positive, the distances in the opposite direction could be regarded as negative.

Thus, if we now say that the man is -2 km from O, we mean that he is 2 km in the direction opposite to the original. Accordingly, in the diagram showing the movements from O (Fig. 8)

Distances to the right with $+$ signs we would call positive.
Distances to the left with $-$ signs we would call negative.

With this device, when giving his position from O, the sign of the number would indicate in which direction the man is from O.

Thus -4 km would indicate he is at C, $+2$ km would show he is at D.

The number with the $+$ sign we call a positive number.
The number with the $-$ sign we call a negative number.

36 Positive and negative numbers

From this it appears that we have devised a new kind of number – viz., a negative number – and that, in consequence, we can divide numbers into two kinds: positive and negative. From the examples above a negative number is a number which in its meaning and effect is opposite to a positive number.

Frequently, as in examples of Sections 32 and 35, the negative number indicates a direction opposite to that of the positive number, and in this sense,

positive and negative numbers are called **directed** numbers.

If negative numbers can rightly be classed as numbers, they must, in operations with them, conform to the rules governing the numbers which we now call positive numbers. These operations will be considered fully later, but a few simple illustrations will serve to show that we can deal with them in the same way as positive numbers.

For example, in the matter of addition, we can add -2 and -3, and a glance at Fig. 8 will show that the result is -5, being equal to the sum of -2 and another -3 from O to the point marked E.

Or if -3 be multiplied by 2 – i.e., we double the distance from O – we get -6, at the point F.

Similarly division of -6 by 2 would give -3.

For the rest of this chapter, in order to make the meaning clear, positive and negative numbers, when being used in operations will be placed in brackets.

Thus $(-6) \div (+2) = (-3)$.

37 Negative numbers

Corresponding to every positive number there is a negative number, and we can write a succession of negative numbers corresponding to positive numbers.

Thus if we write down the numbers beginning, for example, with $+6$ and decreasing by one at each step, we get the series of numbers $+6, +5, +4, +3, +2, +1, 0$.

With the negative number, we do not stop at the zero, but continue with the subtraction, so that we get

$$-1, -2, -3, -4, -5 \ldots, \text{etc.}$$

in descending order of magnitude. Or if we start with (-6) and add unity in succession we get the complete series:

$$-6, -5, -4, -3, -2, -1, 0, +1, +2, +3, +4, +5 \ldots,$$

in ascending order of magnitude. This series can be extended in either direction and decimals and fractions fall into their places between these numbers. Thus we get what is called the **complete number scale**.

38 Graphic representation of the complete number scale

The graphical representation of the complete number scale is so important that we return to it again, using square ruled paper.

Fig. 9.

The straight line XOX' is drawn, as in Fig. 9, to show the representation of a small part of the scale.

On this line, starting from a point O, and using a suitable scale, distances are marked to the right to represent positive numbers and to the left to represent negative numbers.

We could imagine this line to be extended to any distance on either side so that any number could be included. Numbers involving decimals lie between those marked. Thus – 2.5 would be at A.

Two principles may be noted:

(1) Every number can be represented at its appropriate point on the scale.
(2) Conversely, every point on the scale represents a number.

It should be observed that the numbers represented in the figure increase from left to right, as shown by the arrow.

39   Operations with negative numbers

With the introduction of negative numbers, algebra passes beyond the boundaries of arithmetic. We must therefore proceed to examine operations with this new kind of number, remembering, as previously stated, that it must conform to the laws of algebra. The important operations for our present consideration are the fundamental ones of addition, subtraction, multiplication and division.

40   Addition of positive and negative numbers

We have already seen in Section 36 that addition of two negative numbers is performed in the same way as that of positive numbers.

Just as $\qquad (+3) + (+2) = +5$
so $\qquad (-3) + (-2) = -5.$

Such operations can be confirmed by use of Fig. 9.

The addition of a positive and a negative number can also be seen from Fig. 9.

For example, $(-4) + (+3)$ is represented by starting at D, which represents -4, and moving $+3$ to the right to E, the result being -1.

Similarly $(+3) + (-7)$ as found by starting at F, marking $+3$, and since -7 is a negative number, we move 7 divisions to the left to D to find the sum, which is (-4).

When the negative numbers involve letters, the procedure is the same. Thus:

$$(-5a) + (+a) = -4a$$
$$(+2b) + (-5b) = -3b$$
$$(-2x) + (+6x) = 4x.$$

41   Subtraction

This operation presents a little more difficulty, since it is not easy at first to understand what is implied by the subtraction of a negative number, as, for example, $(+6) - (-2)$ or $(-2) - (-5)$.

This can be deduced from Fig. 9, but we will first obtain the rule by applying a fundamental law of addition and subtraction.

Since $\qquad\qquad 9 = 7 + 2$

then $\qquad\qquad 9 - 2 = 7$

and $\qquad\qquad 9 - 7 = 2,$

or, in general terms: if

$$a = b + c$$

then $\qquad\qquad a - b = c$

and $\qquad\qquad a - c = b.$

We have seen above that

$$(-5) + (+3) = -2$$

or $\qquad\qquad\qquad (-2) = (-5) + (+3).$

∴ from the above

$$(-2) - (-5) = +3$$

but we know that $\qquad (-2) + (+5) = +3.$

∴ comparing the two statements

$$-(-5) = (+5).$$

A similar result will clearly hold whatever the numbers.

∴ We conclude that for any number a

$$-(-a) = +a.$$

Similarly $\qquad\qquad -(-2a) = +2a.$

Examples

$$5x - (-3x) = 5x + 3x = 8x$$
$$-2b - (-4b) = -2b + 4b = 2b.$$

Graphical illustrations

To find $(-2) - (-5)$.

The rule can be deduced from the graphical representation of the number scale in Fig. 9 as follows:

If we add a negative number we move to the left along the scale.

Thus $(-2) + (-5) = -7.$

Consequently if we subtract a negative number we must move to the right.

Starting from (-2) and moving 5 to the right we reach $+3$, i.e.,

$$(-2) - (-5) = (+3).$$

To find $(-2) - (+5)$.

When adding a positive number we move to the right.

∴ when subtracting a positive number we move to the left.

∴ starting from (-2) we move 5 divisions to the left and read (-7).

Summarising the rules for addition and subtraction we have:

$$+(+a) = (+a)$$
$$+(-a) = (-a)$$
$$-(+a) = (-a)$$
$$-(-a) = (+a).$$

The student should compare these with rules for signs given in Section 28.

Exercise 8

1   A lift starting from the ground-floor rises to the fourth floor. Then it descends to the second floor, rises to the sixth floor and finally descends to the ground-floor. Express its movements by using positive and negative numbers.

2   The movement of the mercury in a thermometer was as follows. Starting at $+8°C$ it rose $2C°$, fell $14C°$, then rose $4C°$ and finally fell $6C°$. Express these, using positive and negative signs, and find the final temperature.

3   How much higher than a temperature of $-15°C$ is:

   (1) A temperature of $-4°C$?
   (2) Freezing point?
   (3) $+15°C$?

4 Using the number scale shown in Fig. 9 find:

  (1) By how much -2 is greater than -7?
  (2) By how much -6 is less than -1?
  (3) By how much $+3$ is greater than -5?

5 (1) What must be added to (-3) to give (a) -1, (b) $+1$?
  (2) What must be taken from (-3) to give (-8)?

6 Write down the values of:

  (1) $(+6) + (-2)$ (2) $(+6) - (-2)$
  (3) $(-6) + (-2)$ (4) $(-6) - (-2)$
  (5) $0 - (-3)$ (6) $(-4) + (-4)$
  (7) $-(4) - (-4)$ (8) $(-4) - (+4)$

7 Simplify the following:

  (1) $+2a - (-5a)$
  (2) $-4x - (+3x)$
  (3) $+3ab - (-7ab)$
  (4) $2x - 3y - 5y - 3x$
  (5) $(3a - 2b) - (2a + 5b)$
  (6) $(3x - y) - (4x - 3y)$
  (7) $3x - (3y - 4x)$
  (8) $(5 + x) - (6 - 2x) - (3x + 7)$

8 (1) Subtract $(x - 2y)$ from $(3x - 4y)$.
  (2) Subtract $(x - y + 2z)$ from $(3x - 2y - 5z)$.

9 Fill in the brackets in the following:

  (1) $3a - (\quad) = 8a$
  (2) $5x - (\quad) = -x$
  (3) $-3a - (\quad) = 7a$

10 Write down the values of:

  (1) $0 - (+a)$ (2) $0 - (-a)$
  (3) $0 + (-a)$

———————————

42 Multiplication

(1) Multiplication of $(-a)$ by $(+b)$
and $(+a)$ by $(-b)$.

  Consider, as a special case $(-2) \times (+3)$.

Since multiplication is a shortened form of addition, the meaning of $(-2) \times (+3)$ is $(-2) + (-2) + (-2)$.

This $= (-6)$.

$$\therefore \qquad (-2) \times (+3) = (-6).$$

This can be applied to any pair of numbers, and so we may conclude that in general

$$(-a) \times (+b) = (-ab).$$

Since the multiplication of two numbers can be taken in any order (Section 19)

$$(+a) \times (-b) = (-b) \times (+a)$$

and this by the above result is $(-ab)$.

(2) Multiplication of $(-a)$ by $(-b)$.

Since $\qquad (-a) \times (+b) = (-ab)$

and a negative number operates in the opposite sense to a positive number, it follows that

$$(-a) \times (-b) = (+ab).$$

43 Division

(1) Division of $(+a)$ by $(+b)$.

Since $\qquad (+4) \times (+3) = +12,$
$\therefore \qquad (+12) \div (+4) = (+3).$

Similarly $\qquad +(a) \div (+b) = +\left(\dfrac{a}{b}\right).$

(2) Division of $(-a)$ by $(+b)$.

Since $\qquad (+4) \times (-3) = (-12),$
$\therefore \qquad (-12) \div (+4) = (-3).$

Similarly $\qquad (-a) \div (+b) = \left(-\dfrac{a}{b}\right).$

(3) Division of $(+a)$ by $(-b)$.

Again $\qquad (-4) \times (-3) = (+12)$
and $\qquad (+12) \div (-4) = (-3).$

Similarly $\qquad (+a) \div (-b) = \left(-\dfrac{a}{b}\right).$

(4) Division of $(-a)$ by $(-b)$.

As above $(+4) \times (-3) = (-12),$

$\therefore$ $(-12) \div (-3) = (+4),$

and in general $(-a) \div (-b) = \left(+\dfrac{a}{b}\right).$

44 Summary of rules of signs for multiplication and division

Multiplication

$$(+a) \times (+b) = +ab$$
$$(+a) \times (-b) = -ab$$
$$(-a) \times (+b) = -ab$$
$$(-a) \times (-b) = +ab$$

Division

$$(+a) \div (+b) = \left(+\frac{a}{b}\right)$$

$$(+a) \div (-b) = \left(-\frac{a}{b}\right)$$

$$(-a) \div (+b) = \left(-\frac{a}{b}\right)$$

$$(-a) \div (-b) = \left(+\frac{a}{b}\right)$$

These results can be summarised in the following rule:

In the multiplication and division of positive and negative numbers, if the two numbers have the same sign the result is a positive number. If the signs are different, the result is a negative number.

Or to remember these more readily the following slogan can be used:

Like signs give $+$
Unlike signs give $-$

45 Powers, squares and square roots

When we square a number we multiply two numbers with the same signs. In accordance with the above rules, the product must be positive.

Thus
$$(+a) \times (+a) = +a^2$$
$$(-a) \times (-a) = +a^2.$$

Consequently the square of any number is positive.

It follows that when this operation is reversed and the square root of a^2 is required, this may be either $(+a)$ or $(-a)$.

To indicate this we use the sign $\pm$, meaning 'plus or minus', i.e.,

$$\sqrt{a^2} = \pm \, a.$$

Again

$$(-a)^3 = (-a) \times (-a) \times (-a) = -a^3$$

and

$$(-a)^4 = (-a) \times (-a) \times (-a) \times (-a) = +a^4$$

From these and similar examples we may deduce that:

An odd power of a negative number is negative.
An even power of a negative number is positive.

Exercise 9

1 Write down the answers to the following:

(1) $(+12) \times (+3)$ (2) $(+12) \times (-3)$
(3) $(-12) \times (+3)$ (4) $(-12) \times (-3)$
(5) $(+12) \div (+3)$ (6) $(+12) \div (-3)$
(7) $(-12) \div (+3)$ (8) $(-12) \div (-3)$

2 Write down the answers to the following:

(1) $(+a) \times (-a)$ (2) $(-a) \times (-a)$
(3) $(+a) \div (-a)$ (4) $(-a) \div (-a)$

3 Write down the answers to the following:

(1) $(-2a) \times (+2b)$ (2) $(-2a) \times (-2b)$
(3) $(+10x) \times (-2y)$ (4) $(-10x) \times (+2y)$
(5) $(+10x) \div (-2y)$ (6) $(-10x) \div (-2y)$

4 Find the values of the following:

(1) $(-4) \times (+3) \times (-2)$
(2) $(-a) \times (+3a) \times (-2a)$
(3) $(-18xy) \div (-6x)$
(4) $(-24a^2b^2) \div (+4ab)$

5 Find the values of the following:

(1) $(-5x) \times (-2x) \times (-x)$
(2) $a(a - b) - b(b - a)$
(3) $-\{a(-2b) \times (-b)\}$
(4) $-a(a - 2b - c)$

6 Find the simplified form of the following:

(1) $(+2a) \times (-5b) \times (-2b)$
(2) $(-4x)^2 - 2x(5x - 4)$
(3) $x(y - z) - z(x - y) - y(x - z)$

7 Write down the second, third, fourth and fifth powers of:

(1) $(-a)$ (2) $(-2x)$ (3) $\left(-\dfrac{b}{3}\right)$

8 Write down the square roots of 81 and $9x^4$ and the cube roots of $-x^3$ and $-8a^6$.

9 Find the answers to the following:

(1) $(-8x) \times (-2)$
(2) $(-10x) \div (-2)$
(3) $(-2xy) \div (-x)$
(4) $(+6b) \div (-3)$
(5) $(+8t^2) \div (-4t)$
(6) $(-4x^3) \div (-2x^2)$
(7) $(-2x^2) \times (-4x)$
(8) $(+15x^2y) \div (-5xy)$
(9) $(-12a^2b^2) \div (+3ab)$
(10) $(-24a^3bc^2) \div (-4abc)$

10 Write down the values of:

(1) $\{(-a) \times (-b)\} \div (-a)$

(2) $\left(+\dfrac{1}{x^2}\right) \times (-x)^3$

(3) $(-6x)^2 \times (-x)^3 \div (-2x)^4$

5

Expressions and Equations

46 Understanding expressions

Many situations can be described using letters as a shorthand. For example when we find the area of a rectangle we multiply the length by the breadth. If we use 1 to represent the length and b to represent the breadth then the area can be written as $1 \times b$ or more simply by using the expression lb.

In algebra we often use the letter n to represent the position of the nth term in a sequence of numbers. An expression is then given which shows how we can find the value of the term for a particular value of n.

Examples: The nth term in the sequence 2, 4, 6, 8, 10, . . . is 2n. Here the 5th term is 2×5 i.e. 10
The nth term in the sequence 1, 4, 9, 16, 25, . . . is n^2.
Here the 5th term is 5×5 i.e. 25.

Often the expressions will be much more complicated than these and we will clearly need to understand exactly what the expression means before we can use it to find the value of a particular term in the sequence.

We saw on page 19 in chapter 1 how we could use the expression $3n + 1$ to describe the nth term in the sequence of numbers 4, 7, 10, 13 . . . . The 5th number in this sequence is found by substituting 5 as the value of n in the expression, i.e. $3 \times 5 + 1$. The 5th number is therefore 16.

The expression $3n + 1$ means:

first multiply the number n by 3 and then add 1.

A helpful way to illustrate this expression is to draw a diagram.

number → multiply by 3 → add 1 → 3 × number + 1

This can be shortened to:

n → × 3 → + 1 → 3 × n + 1 or 3n + 1

A diagram like this is often called a function machine. The arrows are used to show the order in which the operations are carried out.

If we use a particular value of n, say 5, this is fed through the machine and if necessary we show the intermediate steps in the calculation.

5 → × 3 → 15 → + 1 → 3 × 5 + 1 or 16

47 Using function machines

A function machine is not only a useful aid in helping us to write the correct expression to describe the nth term in a sequence, but also in helping us to carry out the arithmetic operations in the correct order.

Example 1: Draw a function machine to describe:
(a) 5, 8, 11, 14, 17 . . .
     i.e. *first multiply by 3 and then add 2.*
(b) 9, 12, 15, 18, 21 . . .
     i.e. *first add 2 and then multiply by 3.*

  What is the result using 5 with each machine?

(a) n → × 3 →   + 2 → 3 × n + 2 or 3n + 2
    5 → × 3 → 15 → + 2 → 17

(b) n → + 2 →   × 3 → (n + 2) × 3 or 3(n + 2)
    5 → + 2 → 7 → × 3 → 21

You should notice that the order in which the operations are carried out affects both the actual form of the expression and also the numerical results. The brackets are needed in the second expression to show that the whole of n + 2 is multiplied by the 3.

Example 2: Draw a function machine to describe:

(a) 5, 12, 19, 26, 33 . . .

    i.e. *first multiply by 7 and then subtract 2.*

(b) 3, 6, 11, 18, 27 . . .

    i.e. *first square the number and then add 2.*

What is the result using 5 with each machine?

(a) $n \rightarrow \boxed{\times 7} \rightarrow \boxed{-2} \rightarrow 7 \times n - 2$ or $7n - 2$

    $5 \rightarrow \boxed{\times 7} \rightarrow 35 \rightarrow \boxed{-2} \rightarrow 33$

(b) $n \rightarrow \boxed{\text{square}} \rightarrow \boxed{+2} \rightarrow n \times n + 2$ or $n^2 + 2$

    $5 \rightarrow \boxed{\text{square}} \rightarrow 25 \rightarrow \boxed{+2} \rightarrow 27$

Exercise 10

1 Draw a function machine to describe:

    (*a*) *first multiply by 2 and then add 3*
    (*b*) *first multiply by 6 and then subtract 5*
    (*c*) *first add 3 and then multiply by 2*
    (*d*) *first square the number and then subtract 2*

In each of the above find the result of using 7 with the machine.

2 Describe how to find the nth term of the sequence and then draw the function machine for the corresponding expression.

    (*a*) 3, 6, 9, 12, 15. . .   (*d*) 1, 4, 9, 16, 25 . . .
    (*b*) 2, 5, 8, 11, 14. . .   (*e*) 11, 14, 19, 26, 35 . . .
    (*c*) 1, 4, 7, 10, 13. . .   (*f*) 6, 11, 16, 21, 26 . . .

Example 3: Draw a function machine to show how to evaluate the expression: (a) $5n - 2$ (b) $4(n + 5)$ (c) $n^2 + 5$

In each case write down the first five terms of the corresponding sequence:

(a) $n \rightarrow \boxed{\times 5} \rightarrow \boxed{-2} \rightarrow 5 \times n - 2$ or $5n - 2$

    $1 \rightarrow 3, 2 \rightarrow 8, 3 \rightarrow 13, 4 \rightarrow 18, 5 \rightarrow 23$
    so the first five terms of the sequence are 3, 8, 13, 18 and 23

(b) $n \rightarrow \boxed{+5} \rightarrow \boxed{\times 4} \rightarrow (n + 5) \times 4$ or $4(n + 5)$

    $1 \rightarrow 24, 2 \rightarrow 28, 3 \rightarrow 32, 4 \rightarrow 36, 5 \rightarrow 40$
    so the first five terms of the sequence are 24, 28, 32, 36 and 40

(c) $n \rightarrow \boxed{\text{square}} \rightarrow \boxed{+\,5} \rightarrow n \times n + 5$ or $n^2 + 5$

$1 \rightarrow 6, 2 \rightarrow 9, 3 \rightarrow 14, 4 \rightarrow 21, 5 \rightarrow 30$
so the first five terms of the sequence are 6, 9, 14, 21 and 30

Example 4: Draw a function machine to show how to evaluate the expression: (a) $4(5n - 2)$　(b) $4(n + 5) + 3$　(c) $2(n^2 + 5)$

In each case write down the first five terms of the corresponding sequence:

(a) $n \rightarrow \boxed{\times\,5} \rightarrow \boxed{-\,2} \rightarrow \boxed{\times\,4} \rightarrow 4(5n - 2)$

$1 \rightarrow 12, 2 \rightarrow 32, 3 \rightarrow 52, 4 \rightarrow 72, 5 \rightarrow 92$
so the first five terms are 12, 32, 52, 72 and 92

(b) $n \rightarrow \boxed{+\,5} \rightarrow \boxed{\times\,4} \rightarrow \boxed{+\,3} \rightarrow 4(n + 5) + 3$

$1 \rightarrow 27, 2 \rightarrow 31, 3 \rightarrow 35, 4 \rightarrow 39, 5 \rightarrow 43$
so the first five terms of the sequence are 27, 31, 35, 39 and 43

(c) $n \rightarrow \boxed{\text{square}} \rightarrow \boxed{+\,5} \rightarrow \boxed{\times\,2} \rightarrow 2(n^2 + 5)$

$1 \rightarrow 12, 2 \rightarrow 18, 3 \rightarrow 28, 4 \rightarrow 42, 5 \rightarrow 60$
so the first five terms of the sequence are 12, 18, 28, 42 and 60

Exercise 11

1 Draw a function machine to show how to evaluate the expression:
(*a*) $4n - 3$　(*b*) $4(n + 3)$　(*c*) $4n^2$
For each write down the first five terms of the corresponding sequence.

2 Draw a function machine to show how to evaluate the expression:
(*a*) $4(n - 3 + 5)$　(*b*) $4(5n + 3)$　(*c*) $4(n^2 + 7)$
For each write down the first five terms of the corresponding sequence.

3 Draw a function machine to show how to evaluate the expression:
(*a*) $5(n + 3) - 2$　(*b*) $3(4n - 2) + 5$　(*c*) $3(n + 1)^2 + 7$
For each write down the first five terms of the corresponding sequence.

4 Draw a function machine for:
(*a*) $7(n + 1) + 3$　(*b*) $7n + 10$

For each write down the first five terms of the corresponding sequence.

Are the two sequences the same? Why are the expressions equivalent?

In Question 4 of Exercise 11 the two expressions can be shown to be the same by removing the brackets in
$7(n + 1) + 3 = 7n + 7 + 3 = 7n + 10$.
It is not necessary to simplify more complicated expressions like this before evaluating them, provided you can write down the sequence of operations which enables you to work out the value of the expression for a particular value of n. An example comparing the two methods is shown below.

Example 5: Evaluate the expressions $2(3n^2 + 5) - 4$ when $n = 7$
(*a*) by drawing a function machine
(*b*) by simplifying the expression by removing the brackets

(*a*) The function machine is:

$$n \rightarrow \boxed{\text{square}} \rightarrow \boxed{\times 3} \rightarrow \boxed{+ 5} \rightarrow \boxed{\times 2} \rightarrow \boxed{- 4} \rightarrow$$
$$2(3n^2 + 5) - 4$$

with n = 7 we get:

$$7 \rightarrow \boxed{\text{square}} \rightarrow 49 \rightarrow \boxed{\times 3} \rightarrow 147 \rightarrow \boxed{+ 5} \rightarrow 152 \rightarrow$$
$$\boxed{\times 2} \rightarrow 304 \rightarrow \boxed{- 4} \rightarrow 300$$

(*b*) Simplifying the expression we get:

$$2(3n^2 + 5) - 4 = 6n^2 + 10 - 4 = 6n^2 + 6$$

so when n = 7 we have
$6n^2 + 6 = 6 \times 7^2 + 6 = 6 \times 49 + 6 = 294 + 6 = 300$

Exercise 12

1 What expression is evaluated by the function machine?

(*a*) $n \rightarrow \boxed{\times 4} \rightarrow \boxed{- 3} \rightarrow \boxed{\times 5} \rightarrow$????

(*b*) $n \rightarrow \boxed{+ 3} \rightarrow \boxed{\times 5} \rightarrow \boxed{- 2} \rightarrow$????

(*c*) $n \rightarrow \boxed{\times 6} \rightarrow \boxed{- 5} \rightarrow \boxed{\times 2} \rightarrow \boxed{- 7} \rightarrow$????

(*d*) $n \rightarrow \boxed{+ 4} \rightarrow \boxed{\times 3} \rightarrow \boxed{- 5} \rightarrow \boxed{\times 2} \rightarrow$????

2 Draw a function machine to show how to evaluate the expression:

(*a*) $5(4n - 3) + 6$ (*b*) $4(n + 3)^2 - 5$
(*c*) $7(4(n + 3) - 2) - 5$

Find the value of each expression when n = 1.

3 Evaluate the expression $3(2(n + 1) - 5) + 14$ when n = 3

(*a*) by drawing a function machine
(*b*) by simplifying the expression by removing the brackets.

Note: each of the function machines met so far can be described by a rule of the form: $n \rightarrow f(n)$, where $f(n)$ is an expression containing n's.
In Example 5, $n \rightarrow 2(3n^2 + 5) - 4$,
so here $f(n) = 2(3n^2 + 5) - 4$
When n = 7, we found that $7 \rightarrow 300$. We can also show this as $f(7) = 300$.

48 Function notation

Although so far we have used the letter n to represent the position of the nth term in a sequence we could have used any letter to describe the rule. The rule $n \rightarrow 2n + 3$ is the same as the rule $x \rightarrow 2x + 3$. In each case the rule tells us to multiply the number we started with by 2 and then to add 3.

We say that the **image** of n when using this rule is $2n + 3$, whereas the image of x would be $2x + 3$. In either case the image of 5 is $2 \times 5 + 3$ i.e. 13.

Another way of showing the rule $x \rightarrow 2x + 3$ is to use a **mapping** (or arrow) diagram. The rule tells us to *multiply by 2 and then to add 3*

$$1 \rightarrow 5$$
$$2 \rightarrow 7$$
$$3 \rightarrow 9$$
$$4 \rightarrow 11$$
$$5 \rightarrow 13$$

In general we have $x \rightarrow 2x + 3$

The first set of numbers 1, 2, 3, 4, 5 and 6 is called the **domain** of the mapping.
The second set of numbers 5, 7, 9, 11, 13 and 15 is called the **image** set for this mapping.

In order to distinguish between two different rules we often attach a label. We might use the letter f as the label for the rule x → 2x and the letter g as the label for the rule x → x + 3. So f represents the *multiply by 2* rule and g represents the *add 3* rule.

The notation f(x) is used to represent the image of x when using the f rule Here f(x) = 2x, so f(5) = 10 and f(7) = 14.
In a similar way the notation g(x) is used to represent the image of x when using the g rule. Here g(x) = x + 3, so g(5) = 8 and g(7) = 10.

Note: the rule x → 2x + 3 is really made up of the two rules f and g. Here we first use the rule f (*multiply by 2*) and then use the rule g (*add 3*) on the result. Using a function machine this would be:

$$x \to \boxed{\text{use f}} \to f(x) \to \boxed{\text{use g}} \to g[f(x)] \text{ or } gf(x)$$

or $\quad x \to \boxed{\times 2} \to 2(x) \to \boxed{+ 3} \to 2x + 3$

We can use the letters gf to represent the composite rule x → 2x + 3.

Note: in this context, gf means use the f rule first and then the g rule. This is not the same as fg which means use the g rule first and then the f rule.

The corresponding function machine for fg is:

$$x \to \boxed{\text{use g}} \to g(x) \quad \to \boxed{\text{use f}} \to f[g(x)] \text{ or } fg(x)$$

or $\quad x \to \boxed{+ 3} \to x + 3 \to \boxed{\times 2} \to 2(x + 3)$

The study of functions and their properties is an important part of mathematics. However it is not intended in this book to cover the work in any formal way, or to do any more than to introduce some basic ideas.

Examples:
    (1) Find the composite rule fg where f represents x → 3x and g represents x → x − 2. Find also fg(5)
    (2) Find the composite rule gf where f represents x → x + 3 and g represents x → x². Find also gf(5)

(1) fg means use g (*subtract 2*) first and then use f (*multiply by 3*)

$$x \to \boxed{\text{use g}} \to g(x) \quad \to \boxed{\text{use f}} \to f[g(x)] \text{ or } fg(x)$$

so $\quad x \to \boxed{- 2} \to x - 2 \to \boxed{\times 3} \to 3(x - 2)$

hence fg is $x \rightarrow 3(x - 2)$ or $fg(x) = 3(x - 2)$ and
$fg(5) = 3 \times (5 - 2) = 9$

(2) gf means use f (*add 3*) first and then use g (*square*)

$$x \rightarrow \boxed{\text{use f}} \rightarrow f(x) \quad \rightarrow \quad \boxed{\text{use g}} \rightarrow g[f(x)] \text{ or } gf(x)$$

so $x \rightarrow \boxed{+ 3} \rightarrow x + 3 \rightarrow \boxed{\text{square}} \rightarrow (x + 3)^2$

hence gf is $x \rightarrow (x + 3)^2$ or $gf(x) = (x + 3)^2$ and
$gf(5) = (5 + 3)^2 = 64$

Exercise 13

1 Write down in the form of $x \rightarrow \ldots$ the rules:

(*a*) *double and then add 7.* If this rule is f find $f(3)$
(*b*) *add 7 and then double.* If this rule is g find $g(3)$
(*c*) *add 5 and then square.* If this rule is h find $h(3)$
(*d*) *square and then add 5.* If this rule is k find $k(3)$

2 If f is the rule *subtract 3* and g is the rule *multiply by 2* find:

(*a*) $f(5)$ (*b*) $g(9)$ (*c*) the rule fg (*d*) the rule gf

3 Using the rules f, g, h and k from Question 1, find:

(*a*) $gf(3)$ (*b*) $fg(3)$ (*c*) $hk(3)$ (*d*) $kh(4)$
(*e*) the rule kf

The opposite (or **inverse**) of the rule *add 3* is the rule *subtract 3*.
If f is the rule $x \rightarrow x + 4$, then its inverse is the rule $x \rightarrow x - 4$.
In the same way if g is the rule $x \rightarrow 3x$, its inverse is $x \rightarrow x \div 3$.
The inverse of the rule f is called f^{-1}. The inverse of g is called g^{-1}.

Note: if $f(7) = 11$, then $f^{-1}(11) = 7$, also if $g(7) = 21$, then
$g^{-1}(21) = 7$.

Example 1: Find the inverse of the rule $x \rightarrow 3x + 4$.

This rule tells us first to *multiply by 3* and then to *add 4*. To find
the inverse of this rule you need to 'undo' the *add 4* first and then
to 'undo' the *multiply by 3*, i.e. you first *subtract 4* and then *divide
by 3*. So the inverse of $x \rightarrow 3x + 4$ is $x \rightarrow (x - 4) \div 3$.
This process can be seen most easily using a function machine.
The function machine for the rule $x \rightarrow 3x + 4$ is

$$x \rightarrow \boxed{\times 3} \rightarrow 3x \rightarrow \boxed{+ 4} \rightarrow 3x + 4$$

If we want to undo this process we must run the machine 'backwards'.

i.e. $\qquad x \leftarrow \boxed{\div 3} \leftarrow 3x \leftarrow \boxed{-4} \leftarrow 3x + 4$

So reading from right to left we first *subtract 4* and then *divide by 3*. This is the rule $x \rightarrow (x - 4) \div 3$.

Example 2 Find the inverse of the rule $x \rightarrow 3(x + 4) - 2$

The function machine for the rule $x \rightarrow 3(x + 4) - 2$ is

$$x \rightarrow \boxed{+ 4} \rightarrow \boxed{\times 3} \rightarrow \boxed{- 2} \rightarrow 3(x + 4) - 2$$

If we run the machine 'backwards' we get

$$x \leftarrow \boxed{- 4} \leftarrow \boxed{\div 3} \leftarrow \boxed{+ 2} \leftarrow 3(x + 4) - 2$$

So reading from right to left we first *add 2*, then we *divide by 3* and then we *subtract 4*. This gives the rule $x \rightarrow (x + 2) \div 3 - 4$.

You can check that this is correct by trying out a specific value. Say $x = 6$.

Using the rule:

$$x \rightarrow 3(x + 4) - 2, \qquad 6 \rightarrow 3 \times (6 + 4) - 2 \quad = 28$$

Using the inverse rule

$$x \rightarrow (x + 2) \div 3 - 4, 28 \rightarrow (28 + 2) \div 3 - 4 = 6$$

Example 3: Find the number which gives 19 as the result when using the rule $x \rightarrow 3x + 4$

The function machine for the rule $x \rightarrow 3x + 4$ is:

$$x \rightarrow \boxed{\times 3} \rightarrow \boxed{+ 4} \rightarrow 3x + 4$$

We know that the result has to be 19, so we must run the machine 'backwards' to find the original number. In this case it is 5.

i.e. $\qquad 5 \leftarrow \boxed{\div 3} \leftarrow 15 \leftarrow \boxed{- 4} \leftarrow 19$

Exercise 14

1 (*a*) Draw a function machine for the rule $x \rightarrow 4x + 3$
   What is the result if the number used is 5?
   (*b*) Now run the machine backwards to find the inverse of the rule.
   Check that 5 is the result if the number used is 23?

2   (*a*) Draw a function machine for the rule $x \rightarrow 5(x - 4) + 3$
        What is the result if the number used is 8?
    (*b*) Now run the machine backwards to find the inverse of the
        rule.
        Check that 8 is the result if the number used is 23?
3   Find the inverse of the rule $x \rightarrow 2x - 3$
    Use this to find what number gives 11 when using the original
    rule.
4   Find the inverse of the rule $x \rightarrow 5(2x - 3) + 7$
    Hence find the value of x for which $5(2x - 3) + 7 = 22$

49   An introduction to solving equations

The ideas introduced in *Example 3* on page 57 enable us to solve
simple problems of the form: For what value of x is $3x + 4 = 19$?
This is an example of a linear equation which will be considered
in greater detail in the next chapter.

Example 1: For what value of x is $5x - 7 = 8$?

First we need to draw the function machine which gives the
expression on the left-hand side of the equation.

$$x \rightarrow \boxed{\times 5} \rightarrow \qquad \boxed{-7} \rightarrow 5x - 7$$

We want to find the number which gives 8 when using this
machine. To do this we draw the inverse machine using 8 as the
starting number.

$$3 \leftarrow \boxed{\div 5} \leftarrow 15 \leftarrow \boxed{+7} \leftarrow 8$$

This shows that $5x - 7 = 8$ when $x = 3$. To check substitute
$5 \times 3 - 7 = 8$.

Example 2: For what value of x is $5(x - 3) = 20$?

First draw the function machine for the left-hand side of the
equation.

$$x \rightarrow \boxed{-3} \rightarrow \qquad \boxed{\times 5} \rightarrow 5(x - 3)$$

Now draw the inverse machine and use the right-hand number
i.e. 20

$$7 \leftarrow \boxed{+3} \leftarrow 4 \leftarrow \boxed{\div 5} \leftarrow 20$$

So $5(x - 3) = 20$ when $x = 7$. Check: $5(7 - 3) = 5 \times 4 = 20$

Example 3: For what value of n is $3(2n - 5) + 4 = 55$?

First draw the function machine for the left-hand side of the equation.

$$n \rightarrow \boxed{\times 2} \rightarrow \boxed{-5} \rightarrow \boxed{\times 3} \rightarrow \boxed{+4} \rightarrow 3(2n - 5) + 4$$

Now draw the inverse machine and use the right-hand number i.e. 55

$$11 \leftarrow \boxed{\div 2} \leftarrow 22 \leftarrow \boxed{+5} \leftarrow 17 \leftarrow \boxed{\div 3} \leftarrow 51 \leftarrow \boxed{-4} \leftarrow 55$$

So $3(2n - 5) + 4 = 55$ when $n = 11$.
Check: $3(2 \times 11 - 5) + 4 = 3 \times 17 + 4 = 55$

Providing we understand how the expression on the left hand side of the equation is built up and can draw the corresponding function machine, this type of problem is then a matter of working backwards using the inverse machine. Remember to check your answer by substitution.

Exercise 15

1 For what value of x is:

    (*a*) $3x + 7 = 19$ (*b*) $5(x - 4) = 30$ (*c*) $x \div 5 + 3 = 8$

2 For what value of x is:

    (*a*) $7(x + 4) - 3 = 67$ (*b*) $2(3x - 7) + 9 = 25$

3 For what value of n is:

    (*a*) $3(4n - 7) + 5 = 44$ (*b*) $2(5(n + 1) - 3) + 7 = 41$

6

Linear Equations

50   Meaning of an equation

If it is known that 5 times a certain number is 40, a simple process in arithmetic enables us to calculate that the number is 8.

In algebraic form the problem could be expressed as follows:

Let n = the unknown number. Then the question can be put in this way:

If 5n = 40, what is the value of n?

The statement 5n = 40 is called an equation. It is a statement of equality, but it also implies that a value of n is required which will make the left-hand side of the equation equal to the right, or as we say 'satisfies the equation'. The process of finding the value of n which thus satisfies the equation is called 'solving the equation'.

The solution of the above equation involves no more than the division of the right-hand side by the coefficient of n, and could be stated thus:

$$5n = 40$$
$$\therefore \quad n = 40 \div 5$$
$$\therefore \quad n = 8.$$

The solution of an equation is rarely so simple as this. Equations usually consist of more or less complicated expressions on both sides of the equation. By various operations we aim at reducing the equation to the simple form above. The value of the unknown letter is then easily found. These operations will be illustrated in the examples which follow.

51 Solving an equation

Example 1: If 8 times a number is decreased by 5 the result is 123. What is the number?

This simple problem could be solved mentally, but it will serve as an introduction to the process of solution.

Let n = the number.

Then 8n − 5 is the expression which states algebraically '8 times the number decreased by 5'.

But this is equal to 123.

Hence we can form the equation

$$8n - 5 = 123.$$

This is the first step that must always be taken – to formulate the equation. Then we proceed to the solution – i.e., to find the value of n which satisfies it.

Now, the above statement means that 123 is 5 less than 8 times the number, or, if 123 be increased by 5, it is equal to 8 times the number.

∴ we can write the equation in this form

$$8n = 123 + 5.$$

Thus we have practically reached the form we wanted, after which we can readily find the solution.

This step was reached in effect, by transferring the 5 to the right-hand side, leaving only a multiple of n, the unknown number, on the left side. In this transference the argument involved changing the sign of the 5.

The same result could be obtained as follows:

Since $8n - 5 = 123,$

if each side be increased by 5 the result will be that we shall be left with 8n only on the left-hand side and the two sides will still be equal. We shall have as our equation:

$$8n - 5 + 5 = 123 + 5$$
whence $8n = 128$
and $n = 16.$

This device is employed in the solution of practically every equation and it depends on the fact that:

(A) If the same number be added to, or subtracted from, both sides of an equation, the two sides will again be equal.

As a working rule this is equivalent to transferring a number from one side of an equation to the other at the same time changing its sign – i.e., change + to − and − to +.

A principle similar to the above which will be employed later, is:

(B) If both sides of an equation are multiplied or divided by the same number, the two sides of the new equation will be equal. If the multiplier is − 1, both sides change signs.

Example 2: In Section 4 we saw that three consecutive odd numbers could be expressed algebraically by

$$2n + 1, 2n + 3, 2n + 5$$

where n is any integer.

Now suppose we wish to solve this problem:

The sum of three consecutive odd numbers is 81. What are the numbers?

As stated above, the first step is to form an equation. This usually means putting into algebraic form the facts which are given about the unknown number or numbers.

We first, as above, represent the three odd numbers by

$$2n + 1, 2n + 3, 2n + 5.$$

Then we express algebraically the fact that their sum is 81. Hence we get the equation:

$$(2n + 1) + (2n + 3) + (2n + 5) = 81.$$

The use of the brackets helps to make the statement clear.

We now remove the brackets and get:

$$2n + 1 + 2n + 3 + 2n + 5 = 81.$$

Adding like terms $\qquad\qquad 6n + 9 = 81$

whence, as above $\qquad\qquad 6n = 81 - 9$

or $\qquad\qquad\qquad\qquad 6n = 72.$

$\therefore \qquad\qquad\qquad\qquad n = 12.$

We now obtain the odd numbers by substitution of n = 12 in 2n + 1, 2n + 3, 2n + 5.

$\therefore$ the numbers are 25, 27, 29.

This should be checked by ascertaining that their sum is 81.

52 Worked examples

Equations arise out of practical problems in a variety of ways, and examples will be given later, but it is desirable that you should first have sufficient practice in the methods of solving equations. Examples of equations will therefore be worked out and provided for practice which will have no relation to any special problems.

It is usual in such practice equations to use letters at the end of the alphabet, x, y and z, to represent the unknown numbers, and, when necessary, letters at the beginning of the alphabet, a, b, c, etc., to represent known numbers.

This choice of letters is due to Descartes (seventeenth century).

Example 1: Solve the equation:

$$6x - 5 = 2x + 9.$$

The general plan adopted is to collect the terms involving the unknown number, x, on the left side, and the other terms on the right.

Transferring the x term from the right side we get:

$$6x - 2x - 5 = 9.$$

Transferring the -5,

$$6x - 2x = 9 + 5$$
$$\therefore \qquad 4x = 14$$
and
$$x = \frac{7}{2}.$$

Note: With practice the two transference steps could be taken together.

Check. The accuracy of the solution to an equation can always be checked by substituting the value found in both sides of the original equation. In the above case:

Left side $\left(6 \times \dfrac{7}{2}\right) - 5 = 16.$

Right side $\left(2 \times \dfrac{7}{2}\right) + 9 = 16.$

The two sides are equal.

$$\therefore \qquad x = \frac{7}{2} \text{ satisfies the equation.}$$

Example 2: Solve the equation:

$$10(x - 4) = 4(2x - 1) + 5.$$

First simplify both sides by removing brackets.

Then

$$10x - 40 = 8x - 4 + 5.$$

Transferring 8x to the left side and -40 to the right,

$$10x - 8x = 40 - 4 + 5.$$
$$\therefore \qquad 2x = 41$$

and

$$x = 20\tfrac{1}{2}.$$

Check.

Left side

$$10(20\tfrac{1}{2} - 4) = 10 \times 16\tfrac{1}{2} = 165.$$

Right side

$$4(2 \times 20\tfrac{1}{2} - 1) + 5 = 160 + 5 = 165.$$
$$\therefore \qquad x = 20\tfrac{1}{2} \text{ satisfies the equation.}$$

Example 3: Solve the equation:

$$\frac{3x}{5} + \frac{x}{2} = \frac{5x}{4} - 3.$$

When the equation involves fractions, the first step, in general, towards simplification is to 'clear the fractions'. This is effected by multiplying throughout by such a number that the fractions disappear. This is justified by Principle B, Section 51.

The smallest number which will thus clear the fractions is the L.C.M. of their denominators, in this case 20.

Multiplying every term on both sides by 20 we get:

$$\left(\frac{3x}{5} \times 20\right) + \left(\frac{x}{2} \times 20\right) = \left(\frac{5x}{4} \times 20\right) - (3 \times 20).$$
$$\therefore \qquad 12x + 10x = 25x - 60.$$
$$\therefore \qquad 12x + 10x - 25x = -60$$
$$-3x = -60.$$
$$\therefore \qquad x = -60 \div -3.$$
$$\therefore \qquad x = 20 \text{ (by the law of signs).}$$

This should be checked as in previous examples.

Example 4: Solve the equation:

$$4x - \frac{x - 2}{3} = 5 + \frac{2x + 1}{4}.$$

Multiplying throughout by 12

$$48x - 4(x - 2) = 60 + 3(2x + 1). \quad \text{(Section 30)}$$

Clearing brackets

$$48x - 4x + 8 = 60 + 6x + 3.$$
$$\therefore \quad 48x - 4x - 6x = 60 + 3 - 8.$$
$$\therefore \quad 38x = 55$$

and

$$x = \frac{55}{38}.$$

Check by substitution.

Exercise 16

Solve the following equations:

1. (a) $14x = 35$ (b) $1.5x = 30$
2. (a) $\frac{3}{2}x = 24$ (b) $\frac{5}{6}x = 50$
3. (a) $3x = -48$ (b) $-2x = 20$
4. (a) $-\frac{1}{2}x = -40$ (b) $-\frac{1}{3}x = 24$
5. (a) $\dfrac{x}{0.6} = 21$ (b) $\dfrac{2x}{0.2} = -0.8$
6. (a) $5x + 8 = 24$ (b) $10x - 9 = 41$
7. (a) $3x - 4.7 = 2.8$ (b) $2.5x + 50 = 80$
8. (a) $3x - 5 = 2x + 3$ (b) $6y + 11 = 3y + 15$
9. (a) $2x + 6 = 14 - 3x$ (b) $z + 20 = 5z - 44$
10. (a) $3y - 1.5 = 7y - 8.7$ (b) $4.8x + 52 = 3.2x - 20$
11. (a) $3y - \dfrac{44}{5} = 2y + 6$ (b) $x + \dfrac{x}{15} = 20$
12. (a) $\dfrac{3x}{5} - \dfrac{2}{3} = 0$ (b) $\dfrac{x}{0.1} + \dfrac{x}{0.2} = 5$
13. $4(2x - 5) = 3(2x + 8)$
14. $3x - 2(x + 4) = 5x - 28$
15. $2(x + 5) - 3(x - 6) = 20$
16. $5(y - 1) - 2(y + 6) = 2y + 12$
17. $2(x - 1) + 3(x + 4) = 4(x + 1) - (x - 5)$
18. $3(x - 7) - 3(2x - 4) = 4(x + 3)$
19. $\frac{1}{2}(3y + 6) - \frac{1}{3}(2y - 4) = 20$
20. (a) $\dfrac{x}{8} - \dfrac{4}{5} = 0$ (b) $\dfrac{2x}{5} - 3 = 8$
21. (a) $\dfrac{7x}{8} - (x - 2) = 12$ (b) $\dfrac{3x}{2} - \dfrac{x}{3} = \dfrac{5(x - 4)}{6}.$

22 $\dfrac{2x - 5}{3} - \dfrac{3x - 1}{4} = \dfrac{3}{2}$.

23 $\dfrac{7x - 6}{4} - \dfrac{5x + 3}{7} - \dfrac{6x - 1}{14} = 3$

24 $\frac{1}{2}(x + 3) - \frac{2}{5}(x - 3) = x + 9$

25 $12(5 - x) - 3(3x - 4) = 23$

26 $\frac{3}{2}(x - 1) - \frac{1}{3}(5 - x) = 2\frac{1}{3} - 3(x - 3)$

27 $\dfrac{y}{7} - \dfrac{1 - y}{5} = 2 - \dfrac{3y}{5}$

28 $\dfrac{2x + 4}{1.5} - \dfrac{3x - 5}{2.5} = 1$

29 Solve for n the equation $2n = 0.58(12 - n)$.

30 For what value of r is 18.4 equal to $2(3.5r - 1)$?

31 Find x when $\dfrac{7.5}{x} = \dfrac{5}{2}$.

32 Find c if $\dfrac{18}{2c} = 3.8$.

33 If $C = \dfrac{V}{R}$, find V when $C = 8$, $R = 4.5$.

34 For what value of x is $3(x - 5)$ equal to $\dfrac{4x + 3}{2}$?

---

53 Problems leading to simple equations

The methods of solving problems by means of simple equations are illustrated by the following examples. The general method of procedure is:

(1) Having decided which is the unknown quantity, represent it by a letter, such as x, stating clearly the units employed when necessary.
(2) Form an equation which represents the facts provided by the problem about the unknown quantity.
(3) Solve the equation.

Example 1: At a dance party there were 10 more women than men. The men paid 20p each, the women 15p each, and the total receipts were £16.20. How many men and women were there at the party?

There are two unknown quantities: the number of men and the number of women. But if the number of men is known, the number of women is 10 more.

∴   Let x = the number of men.
Then x + 10 = the number of women.
The facts supplied are represented as follows:

$$20x = \text{the money paid by the men, in pence;}$$
$$15(x + 10) = \text{the money paid by the women, in pence;}$$
$$1620 = \text{the total amount paid, in pence.}$$

The equation which connects these is consequently:

$$20x + 15(x + 10) = 1620.$$
∴ $$20x + 15x + 150 = 1620$$
$$35x = 1620 - 150$$
$$35x = 1470.$$
∴ $$x = 42.$$
∴ the number of women is x + 10 = 52.
∴ the solution is 42 men, 52 women.

Example 2: A motorist travels from town A to town B at an average speed of 64 km/h. On his return journey his average speed is 80 km/h. He takes 9 hours for the double journey (not including stops). How far is it from A to B?

The unknown quantity is distance from A to B.
Let x = distance in kilometres.
Now distance = speed × time.

∴ $$\frac{\text{distance}}{\text{speed}} = \text{time.}$$

Then   time for 1st journey $= \frac{x}{64}$.

Then   time for 2nd journey $= \frac{x}{80}$.

But total time is 9 hours.
∴ the equation is

$$\frac{x}{64} + \frac{x}{80} = 9.$$

To clear of fractions multiply throughout by 320.

∴ $$\left(\frac{x}{64} \times 320\right) + \left(\frac{x}{80} \times 320\right) = 9 \times 320$$
$$5x + 4x = 9 \times 320$$
∴ $$9x = 9 \times 320$$
and $$x = 320 \text{ km.}$$

This should be checked by applying the conditions in the question.

Example 3: The weekly wages of a man and woman engaged on the same kind of work were £4.80 and £3.70 respectively. It was agreed to increase the two wages by the same amount so that the man's wage was $\frac{6}{5}$ of the woman's wage. What increase was given?

The unknown quantity is the money to be added to the wages. Let x = the amount of increase in pence.

Then   (480 + x) pence = the man's new wages.
        (370 + x) pence = the woman's new wages.
Then by the data

$$480 + x = \tfrac{6}{5}(370 + x).$$

Clearing fractions

	$5(480 + x) = 6(370 + x).$
∴	$2400 + 5x = 2220 + 6x.$
∴	$2400 - 2220 = 6x - 5x$
and	$180 = x.$

∴ the weekly increase is 180p or £1.80.

This should be checked by adding it to each of the weekly wages and ascertaining if one is $\frac{6}{5}$ of the other.

Exercise 17

1 From three times a certain number, n, 6 is subtracted. The result is equal to twice the number together with 6. What is the value of n?
2 There is a number such that when it is multiplied by 5 and then 14 is subtracted, the result is 348.5. Find the number.
3 From 5 times a certain number 189 is subtracted, and the remainder is one half the original number. What is that number?
4 One-third of a number added to four-fifths of itself is equal to 17. What is the number?
5 When 9 is subtracted from 6 times a certain number, the result is 45 more than twice the number. Find the number.
6 The sum of three consecutive odd numbers is 69. What are the numbers?

7 A man walks from one town to another at an average speed of 2.5 km/h. On the return he quickens his average speed to 3 km/h. The time taken for the double journey was 7 h 20 min. How far are the two towns apart?

8 The sum of a number and 4 per cent of itself is 41.6. What is the number?

9 The perimeter of a rectangle is 44 cm. If one of the two adjacent sides is 1.8 cm longer than the other, what are the lengths of the sides?

10 Some men agree to pay equally for the use of a boat, and each pays 15p. If there had been two more men in the party, each would have paid 10p. How many men were there and how much was the hire of the boat?

11 A man distributes £2 among 20 boys, giving 5p each to some and 25p to the rest. How many boys received 25p each?

12 A man is four times as old as his son. In four years' time he will be three times as old. What are their ages now?

13 The connection between the degrees on the Celsius and Fahrenheit thermometers is that $x°C = \left(\frac{9x}{5} + 32\right) °F$.

What number of degrees Celsius is equivalent to 86°F?

14 A bookseller buys 120 volumes of a certain series of books. He sells some at the published price of 18p each and disposes of the remainder during a sale at 12p each. If his total receipts are £19.20, find how many volumes were sold at each price.

15 A bus is carrying 32 passengers, some with 3p tickets and the remainder with 5p tickets. If the total receipts from these passengers are £1.14, find the number of 3p fares.

7

Formulae

54 Practical importance of formulae

One of the most important applications of elementary algebra is to the use of formulae. In every form of applied science and mathematics, such as mechanical engineering, electrical engineering, aeroplane construction, etc., formulae are constantly employed, and their interpretation and manipulation are essential.

55 Treatment of formulae

In the first chapter of this book some very easy examples of formulae were introduced. With the assistance of a greater knowledge of algebraical symbols and operations and with increasing skill in their use, the student can now proceed to more difficult types.

Formulae involve three operations:

(1) Construction;   (2) manipulation;   (3) evaluation.

The construction of formulae cannot be indicated by any specified rules or methods. A knowledge of the principles of algebra and skill in their application are necessary. But in general you are concerned with formulae which have already been evolved. What you need is skill in using them; and as your knowledge of Algebra increases, so you will be better equipped for dealing with new examples.

The manipulation and evaluation of formulae are closely associated. A formula may need to be re-arranged or simplified before any substitution of values may be made. Experience alone will guide you as to what manipulation is desirable in order to reach a form which is the most suitable for evaluation or some other purpose. Clear arrangement of working is always essential for accuracy.

56 Worked examples

Example 1: Find a formula for the total area (A) of the surface of a square pyramid as in Fig. 10 when AB = a and OQ = d.

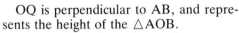

OQ is perpendicular to AB, and represents the height of the △AOB.

The total area is made up of:

(1) Area of base.
(2) Areas of the four △s, of which AOB is one.

Area of base = a^2.
Area of each △ = $\frac{1}{2}$ad.
∴ area of all △s = 4 × $\frac{1}{2}$ad = 2ad.
∴ total area of pyramid = a^2 + 2ad

Fig. 10.

or $\qquad$ A = a(a + 2d).

Example 2: If $L = \dfrac{W(T - t)}{w} - t$,

find L when t = 8.5, w = 115, W = 380, T = 28.5.

Substituting in $\quad L = \dfrac{W(T - t)}{w} - t$.

$$L = \frac{380(28.5 - 8.5)}{115} - 8.5$$

$$= \frac{380 \times 20}{115} - 8.5$$

$$= 66 - 8.5 \text{ (approx.)}$$

$$= 57.5 \text{ (approx.)}$$

Exercise 18

1 If $s = ut + \frac{1}{2}at^2$, find s when u = 15, t = 5, a = 8.
2 The volume of a cone, V, is given by the formula
$V = \frac{1}{3}\pi r^2 h$, where r = radius of base, h = height of cone.
Find V when r = 3.5, h = 12, $\pi = \dfrac{22}{7}$.

3   The volume of a sphere is given by the formula

   $V = \frac{4}{3}\pi r^3$, where r = radius. Find V when r = 3, $\pi = \frac{22}{7}$.

4   If $E = \frac{Wv^2}{2g}$, find E when W = 15.5, v = 18.8, and g = 32.

5   From the formula $C = \frac{E + e}{R + r}$, find C when E = 17.6,

   e = 1.5, R = 28.4, r = 2.6.

6   In a suspension bridge the length of the cable employed is

   given by the formula $L = 1 + \frac{8d^2}{3l}$, where

> L = length of the cable,
> d = dip of the centre of the cable,
> 1 = length of the span of the bridge,

   all measurements being in metres. Find L when d = 6, l = 56.

7   A formula for the loading of beams is $W = \frac{4kbd^2}{l}$.

   Find W when k = 45, b = 2, d = $\frac{1}{2}$, l = 20.

8   If $s = \frac{n(n + 1)(2n + 1)}{6}$, find s when n = 8.

9   The following formula is used in connection with pile driving,

   $L = \frac{Wh}{d(W + P)}$. Find L when W = 5, d = 1.5, P = 19,

   h = 4.5.

10   If R = W(x + 3t), find R when W = 210, x = 6.5, t = 0.04.

11   From the formula $H = \frac{plAN}{33\ 000}$ find H when p = 18, l = 2,

   A = 80, N = 360.

---

57   Transformation of formulae

In the formulae which have been examined it will be seen that one quantity is expressed in terms of other quantities and the formula expresses the relations between them. Thus in the formula for the volume of a cone ($V = \frac{1}{3}\pi r^2 h$, Exercise 18, No. 2) this volume is expressed in terms of the height and the radius of the base.

But it may be necessary to express the height of the cone in terms of the volume and the radius of the base. In that case we would write the formula in the form:

$$h = \frac{3V}{\pi r^2} ,$$

that is, the formula has been transformed.

When one quantity is expressed in terms of others, as in $V = \frac{1}{3}\pi r^2 h$, the quantity thus expressed, in this case V, is sometimes called the subject of the formula.

When the formula was transformed into

$$h = \frac{3V}{\pi r^2},$$

the subject of the formula is now h. This process of transformation has been termed by Prof. Sir Percy Nunn 'changing the subject of the formula'.

The transformatin of formulae often requires skill and experience in algebraical manipulation; the following examples will help to illustrate the methods to be followed.

58 Worked examples

Example 1: From the formula $T = \frac{\pi f d^3}{16}$, find:

(1) f in terms of the other quantities.
(2) d in terms of the other quantities.

From
$$T = \frac{\pi f d^3}{16},$$

clearing the fraction $\quad 16T = \pi f d^3 \qquad\qquad (1)$

or $\qquad\qquad \pi f d^3 = 16T,$

dividing throughout by πd^3 $\quad f = \frac{16T}{\pi d^3}.$

From (1) dividing throughout by πf

$$d^3 = \frac{16T}{\pi f}.$$

$\therefore \qquad\qquad d = \sqrt[3]{\frac{16T}{\pi f}}.$

Example 2: Transform the formula
$$L = 1 + \frac{8d^2}{3l}$$
into one which expresses d in terms of the other quantities.

$$L = 1 + \frac{8d^2}{3l}.$$

Clearing fractions: $3lL = 3l^2 + 8d^2$

or $d^2 = 3lL - 3l^2$.

$\therefore$ $d^2 = \dfrac{3lL - 3l^2}{8}$.

$\therefore$ $d = \sqrt{\dfrac{3lL - 3l^2}{8}}$.

Example 3: The velocity, V, of water flowing through a pipe, occurs in the formula,

$$h = 0.03 \,\frac{L}{D} \times \frac{V^2}{2g} .$$

Change the subject of the formula to V.

Writing the formula as

$$0.03 \,\frac{L}{D} \times \frac{V^2}{2g} = h.$$

Divide both sides by $0.03 \,\dfrac{L}{D}$,

$$\frac{V^2}{2g} = h \div \frac{0.03L}{D}$$

$$= h \times \frac{D}{0.03L}$$

$$= \frac{hD}{0.03L} .$$

$\therefore$ $V^2 = \dfrac{2ghD}{0.03L}$.

$\therefore$ $V = \sqrt{\dfrac{2ghD}{0.03L}}$.

Example 4: If $a - b = x(c - nd)$ find n in terms of the other letters.

Attention should be fixed on the term containing n, viz. nd. This we try to isolate from the other terms.

$\therefore$ divide both sides by x.

Then $\dfrac{a - b}{x} = c - nd$.

Transferring c

$$\frac{a - b}{x} - c = -nd .$$

$$\therefore \qquad nd = c - \frac{a - b}{x} ,$$

$$\therefore \qquad n = \frac{1}{d} \left\{ c - \frac{a - b}{x} \right\} .$$

Example 5: The time of vibration of a simple pendulum is given by the formula

$$t = 2\pi \sqrt{\frac{l}{g}} .$$

Find l in terms of the other quantities.

$$t = 2\pi \sqrt{\frac{l}{g}} .$$

Square both sides.

Then
$$t^2 = 4\pi^2 \times \frac{l}{g} .$$

$$\therefore \qquad t^2 g = 4\pi^2 l$$

or
$$4\pi^2 l = gt^2.$$

$$\therefore \qquad l = \frac{gt^2}{4\pi^2} .$$

Exercise 19

1 The formula for the area (A) of a circle, in terms of its radius (r) is $A = \pi r^2$. Change the subject of the formula to express r in terms of the area.

2 Transform the formula for the volume of a sphere – viz. $V = \frac{4}{3}\pi r^3$ (see Exercise 18, No. 3) – into a formula in which r is expressed in terms of the volume.

3 Change the formula for the volume of a cone – viz., $V = \frac{1}{3}\pi r^2 h$ – to a formula in which the subject is r.

4 The horse power of a motor is given by the formula $H = \frac{EC}{825}$. Express this as a formula for C.

5 The lifting force of an electro-magnet is given by the formula $F = \frac{B^2 A}{112 \times 10^5}$, where F is the force.

Transform this into a formula of which the subject is B.

6 The amount of sag, d, in a beam under certain conditions is given by the formula $d = \frac{Wl^3}{48EI}$.

Express this as a formula expressing l in terms of the other quantities.

7 If $v^2 = u^2 + 2as$, express s in terms of u, v, and a. Find the value of s when $u = 15$, $v = 20$, and $a = 5$.

8 There is an electrical formula $I = \dfrac{V}{R}$.

Express this (1) as a formula for V and (2) as a formula for R. Find I if $V = 2$ and $R = 20$.

9 If $n^2r + 1 = NR$, rearrange the expression so that it becomes a formula for n. Find the value of n when $N = 25$, $R = 2$, $r = 0.81$.

10 The relation of the volume (v) of a mass of gas to the pressure (p) on it is given by the law $pv = k$.

In a certain experiment when $p = 84$, $v = 12$. Find the value of k and then express the formula giving v in terms of p and the value of k.

---

59 Literal equations

The operations employed in changing the subject of a formula are the same in principle as those used in the solution of equations. One essential difference from the equations dealt with in Chapter 6 is that whereas these were concerned with obtaining numerical values when solving the equations in the formula the quantity which is the subject of the formula is expressed in terms of other quantities, and its numerical value is not determined, except when the numerical values of these quantities are known.

It is frequently necessary, however, to solve equations in which the values of the unknown quantities will be found in terms of letters which occur in the equation. Such equations are termed literal equations. The methods of solution are the same in principle as those employed in Chapter 6. They are illustrated in the following examples.

60 Worked examples

Example 1: Solve the equation $5x - a = 2x - b$.

As pointed out previously (Section 52), x is undestood as standing for the unknown quantity and the use of the letters a and b marks the difference between this kind of equation and those of Chapter 6. The methods by which the equation is solved are the same, however, a and b being treated in the same way as ordinary numerals.

$$5x - a = 2x - b$$

transferring 2x and −a respectively from one side to the other, changing the signs in so doing,

$$5x - 2x = a - b.$$
∴ $$3x = a - b$$
and $$x = \frac{a - b}{3}.$$

Example 2: Solve for x
$$a(x - 2) = 5x - (a + b).$$

Removing brackets

$$ax - 2a = 5x - a - b.$$

Transferring $$ax - 5x = 2a - a - b$$
or $$ax - 5x = a - b.$$

This introduces a point of difference from numerical equations. With the latter we add the terms involving x by adding their coefficients. The addition in this case cannot, however, be made arithmetically. Algebraically the sum of the coefficients of x is

$$a + (-5) \text{ or } a - 5.$$
∴ we write $$(a - 5)x = a - b.$$

Dividing both sides by the coefficient of x

$$x = \frac{a - b}{a - 5}.$$

Exercise 20

Solve the following equations for x:

1 $5x - 4a = 0$
2 $5x - 3a = 7a$
3 $8x - p = 3x + 4p$
4 $3x + 2b = 2(x + 3b)$
5 $ax + b = 3a - b$
6 $b(x - p) = c$
7 $2a - b = b - bx$
8 $3(ax - 2) + 25b = 6b$
9 $p(x - q) = x(p - q)$

10 $\dfrac{ax}{2} - 4 = \dfrac{b}{3} - 6$.

11 $ax - 4b = bx - b$

12 $\dfrac{x}{a} + \dfrac{x}{3a} = 4 - x$.

13 $a(x - a) = b(x + b)$

14 $\dfrac{x}{m} - \dfrac{1}{n} = \dfrac{3x}{mn}$.

8

Simultaneous Equations

61 Simple equations with two unknown quantities

The simple equations considered in Chapter 6 contained only one
unknown quantity whose value it was required to determine. But
many of the formulae quoted in the previous chapter contain
several quantities. Cases may therefore occur in which it will be
required to find the value of more than one of these. Similarly
problems arise which, for their solution, involve the determination
of more than one unknown.

62 Solution of simultaneous equations

A simple problem will serve to illustrate the above statement.
Suppose we are told

The sum of two numbers is 10: what are the numbers?

Let the two numbers be represented by x and y.
Then we know that

$$x + y = 10.$$

It is evident that there is an infinite number of solutions of this
equation, such as (1, 9), (2, 8), (3.5, 6.5), etc.
If the equation be written in the form:

$$x = 10 - y.$$

This gives   x in terms of y.

Whatever value be given to y in this equation, a corresponding value of x can be found, and each pair of values furnishes a solution of the equation.

If a second condition has to be fulfilled we can determine which of these pairs satisfies it.

For example, if, in addition to the statement that the sum of the numbers is 10, we are also told that one of them is four times the other; then there is only one set of the pairs of values referred to above which will satisfy both the conditions.

For if $\qquad\qquad\qquad$ $x + y = 10$ $\qquad\qquad\qquad$ (1)
and $\qquad\qquad\qquad\qquad$ $x = 4y$ $\qquad\qquad\qquad\quad$ (2)

substituting for x in equation (1) we get:

$$4y + y = 10$$
$$5y = 10$$
$$y = 2,$$

and since $\qquad\qquad\qquad\qquad$ $x = 4y,$
then $\qquad\qquad\qquad\qquad\quad$ $x = 8.$

∴ the solution which satisfies both equations simultaneously is $x = 8$, $y = 2$, and clearly there is no other solution.

For this reason such equations are called simultaneous equations.

It is evident that if there are two unknown quantities whose values are required, it must be possible to form two separate equations connecting them.

The methods employed in solving these equations are shown in the following examples.

63 Worked examples

Example 1: Solve the equations:

$$2x + y = 21 \qquad\qquad (1)$$
$$3x + 4y = 44 \qquad\qquad (2)$$

In the method employed in this example we begin by obtaining one letter in terms of the other. The more convenient one is chosen, and in this case from equation (1) $2x + y = 21$, we get:

$$y = 21 - 2x \qquad\qquad (3)$$

We could have found x in terms of y, but this would involve fractions and is not so convenient.

Substituting in equation (2) the value of y thus obtained from equation (1)

$$3x + 4(21 - 2x) = 44.$$

Thus we reach a simple equation with one unknown. This is solved as previously:

$$3x + 84 - 8x = 44$$
$$3x - 8x = 44 - 84$$
$$-5x = -40$$

or
$$5x = 40$$

and
$$x = 8.$$

Substituting this value of x in equation (3) we can find y.

Thus
$$y = 21 - (2 \times 8)$$
∴
$$y = 5.$$

∴ the solution is x = 8, y = 5.

These values should be checked by substitution in both of the given equations.

Example 2: In the following example a second method is shown which can frequently be employed to advantage.

Solve the equations:

$$x + y = 15 \tag{1}$$
$$3x - y = 21 \tag{2}$$

It will be seen that if the left sides of the two equations were added, y would be eliminated, since we would get $(+y) + (-y) = 0$. Thus only x would remain.

It is clear also that the sum of the two left sides of the equations must equal the sum of the two right sides, i.e.,

$$(x + y) + (3x - y) = 15 + 21,$$

whence
$$x + 3x = 36$$
$$4x = 36.$$
∴
$$x = 9.$$

Substituting this value of x in equation (2) or, if easier, in equation (1)

$$(3 \times 9) - y = 21.$$
∴
$$27 - y = 21$$
$$-y = -6.$$
∴
$$y = 6.$$

∴ the solution is x = 9, y = 6.

Example 3: In the following example both of the above methods are employed.

Solve the equations:

$$2x + 3y = 42 \qquad (1)$$
$$5x - y = 20 \qquad (2)$$

1st method. Substitution.

From (2) $\qquad\qquad -y = 20 - 5x.$
∴ $\qquad\qquad\qquad\quad y = 5x - 20 \qquad (3)$

Substituting in (1):

$$2x + 3(5x - 20) = 42$$
$$2x + 15x - 60 = 42.$$
∴ $\qquad\qquad\qquad 17x = 102$
and $\qquad\qquad\qquad x = 6.$

Substituting for x in (3):

$$y = (5 \times 6) - 20$$
and $\qquad\qquad\quad y = 10.$

∴ the solution is x = 6, y = 10.

2nd method. Elimination.

In this example neither letter can be eliminated by addition of the left sides of the equation, as in example 2. But by multiplying both sides of equation (2) by 3, y can be eliminated.

We proceed thus:

$$2x + 3y = 42 \qquad (1)$$
$$5x - y = 20 \qquad (2)$$

Multiplying by 3 throughout in (2)

$$15x - 3y = 60 \qquad (3)$$

Adding (1) and (3)

$$(2x + 3y) + (15x - 3y) = 42 + 60$$
∴ $\qquad\qquad\qquad\qquad 17x = 102$
and $\qquad\qquad\qquad\qquad x = 6.$

From this y can be found as before.

Note: x could have been eliminated from (1) and (2) as follows:

(*a*) Multiply throughout in (1) by 5.
(*b*) Multiply throughout in (2) by 2.

Then we get:
$$10x + 15y = 210$$
$$10x - 2y = 40$$

Subtracting
$$(10x + 15y) - (10x - 2y) = 210 - 40.$$
$$\therefore \quad 10x + 15y - 10x + 2y = 170$$
whence
$$17y = 170$$
and
$$y = 10.$$

Comparison of the methods. Of these two methods, that of substitution is the sounder and more general. In practice equations are seldom easily dealt with by the elimination method.

Example 4: Find values of R_1 and R_2 which will satisfy the equations:

$$0.5R_1 + 1.2R_2 = 1.486 \tag{1}$$
$$4.5R_1 - 2R_2 = 4.67 \tag{2}$$

From (1)

$$0.5R_1 = 1.486 - 1.2R_2.$$
$$\therefore \quad R_1 = \frac{1.486 - 1.2R_2}{0.5}$$
$$= 2.972 - 2.4R_2.$$

Substituting in (2):

$$4.5(2.972 - 2.4R_2) - 2R_2 = 4.67$$
$$13.374 - 10.8R_2 - 2R_2 = 4.67$$
$$\therefore \quad -12.8R_2 = -8.704$$
$$R_2 = \frac{8.704}{12.8}$$
$$\therefore \quad R_2 = 0.68.$$

Substituting in (2):

$$4.5R_1 - (2 \times 0.68) = 4.67$$
$$4.5R_1 - 1.36 = 4.67$$
$$4.5R_1 = 6.03.$$
$$\therefore \quad R_1 = \frac{6.03}{4.5}$$
$$= 1.34.$$

$\therefore$ the solution is $R_1 = 1.34$, $R_2 = 0.68$.

Exercise 21

Solve the following equations:

1 $y = 2x$
$3x + 2y = 21$

2 $y = 3x - 7$
$5x - 3y = 1$

3 $x = 5y - 3$
$3x - 8y = 12$

4 $x - y = 5$
$4x - y = 2x + 13$

5 $3x - 2y = 7$
$x + 2y = 5$

6 $2x - y = 3$
$x + 2y = 14$

7 $2x - y = 10$
$3x + 2y = 29$

8 $2(x - 4) = 3(y - 3)$
$y - 2x = -13$

9 $3x - 2(y + 3) = 2$

$2(x - 3) + 4 = 3y - 5$

10 $\dfrac{y}{2} - x = 2$

$6x - \dfrac{3y}{2} = 3$

11 $\dfrac{x}{2} - \dfrac{y}{5} = 1$

$y - \dfrac{x}{3} = 8$

12 $\dfrac{a}{2} + b = 1$

$3a - \dfrac{b}{3} = \dfrac{31}{2}$

13 $\dfrac{b}{3} - 2a = 5$

$3a + 4b = 6$

14 $4(1 - p) = 7q + 8p$

$6p + q + 8 = 0$

15 $7x + 3(y - 3) = 5(x + y)$
$7(x - 1) - 6y = 5(x - y)$

16 $\dfrac{x}{2} - \dfrac{y}{3} = \dfrac{1}{6}$

$\dfrac{y}{2} - \dfrac{x}{6} = 5$

17 $2(3a - b) = 5(a - 2)$

$3(a + 4b) = 2(b - 3)$

18 $0.1x + 0.2y = -0.2$
$1.5x - 0.4y = 10.6$

19 $1.25x - 0.75y = 1$
$0.25x + 1.25y = 17$

20 $2.5x + 3.7y = 13.365$

$8.2x - 1.5y = 7.02$

21 $\dfrac{x - 1}{3} + \dfrac{y + 2}{2} = 3$

$\dfrac{1 - x}{6} - \dfrac{y - 4}{2} = \dfrac{1}{2}$

22 $2P - 5Q = 2$
$3P + 10Q = 8.6$

23 $3(x - y) - \dfrac{1}{3}(x + y) = 30$

$x + y + \dfrac{5}{3}(x - y) = 22$

24 $\dfrac{x}{3} + \dfrac{y}{2} = \dfrac{2x}{3} - \dfrac{y}{6} = 7$

64   Problems leading to simultaneous equations

Many problems require for their solution the determination of two unknown quantities. The general method of solution is similar to that employed when there is one unknown, but with the important difference that when there are two unknowns to be determined, two equations must be formed from the data.

The following examples illustrate the methods employed.

65   Worked examples

Example 1: There are two numbers such that the sum of the first and three times the second is 53, while the difference between 4 times the first and twice the second is 2. Find the numbers.

Let x = one number.
Let y = the second.
Then from the first set of facts

$$x + 3y = 53 \tag{1}$$

From the second

$$4x - 2y = 2 \tag{2}$$

From (1) $x = 53 - 3y.$
Substituting in (2):

$$4(53 - 3y) - 2y = 2.$$
$\therefore$ $212 - 12y - 2y = 2$
Collecting $- 14y = -210.$
$\therefore$ $y = \dfrac{-210}{-14}$
and $y = 15.$

Substituting in $x = 53 - 3y$
               $x = 53 - (3 \times 15).$
$\therefore$ $x = 8.$

$\therefore$ the numbers are 8 and 15.

Example 2: In the equation $y = mx + b$ it is known that the equation is satisfied by two pairs of values of x and y – viz. when

               $x = 4, y = 6,$
and when $x = 2.4, y = 4.5.$

What are the values of m and b?

This is an example of an important practical problem. It means that there is a law connecting x and y, the law involving m and b, which are constants. These constants must be determined before the law can be stated.

They are therefore the unknown numbers in this case. The equations connecting them are obtained by substituting the given pairs of values of x and y.

(1) When x = 4, y = 6. ∴ on substitution

$$6 = 4m + b \tag{1}$$

(2) When x = 2.4, y = 4.5.

∴ $$4.5 = 2.4m + b \tag{2}$$

These are to be solved simultaneously for a and b. It is clearly a case for using the method of elimination.

Subtracting (2) from (1):

$$6 - 4.5 = (4m + b) - (2.4m + b)$$

or $$6 - 4.5 = 4m + b - 2.4m - b$$

∴ $$1.5 = 1.6m.$$

∴ $$m = \frac{1.5}{1.6} \text{ or } \frac{15}{16} .$$

Substituting in (1)

$$\left(4 \times \frac{15}{16}\right) + b = 6.$$

∴ $$b = 6 - \frac{15}{4}$$

and $$b = \frac{9}{4} .$$

∴ the solution is $$m = \frac{15}{16}, \ b = \frac{9}{4} .$$

Substituting these in

$$y = mx + b$$

we get $$y = \frac{15}{16} x + \frac{9}{4}$$

or clearing fractions $$16y = 15x + 36.$$

Example 3: A bookseller has a number of books the published price of which is 25p. After selling a certain number at this price he sells the remainder at 20p each, and his total receipts are £11. If the numbers sold at the two prices were reversed, he would have received £11.50. How many books had he in all and how many were originally sold at 25p?

Let x = the number originally sold at 25p;
Let y = the number originally sold at 20p.
The amounts received for these were 25x pence and 20y pence and their total value was £11 or 1100 pence.

$\therefore$ $$25x + 20y = 1100.$$

When the numbers are reversed he receives 20x and 25y pence and their total value is now £11.50 or 1150 pence.

$\therefore$ $$20x + 25y = 1150$$

$\therefore$ the equations to be solved simultaneously are:

$$25x + 20y = 1100 \qquad (1)$$
$$20x + 25y = 1150 \qquad (2)$$

From (1) $$25x = 1100 - 20y$$
and $$x = \frac{1100 - 20y}{25}$$

Substituting in (2)

$$\frac{20(1100 - 20y)}{25} + 25y = 1500$$

$\therefore$ $$4(220 - 4y) + 25y = 1150$$
and $$880 - 16y + 25y = 1500$$
$\therefore$ $$9y = 270$$
$$y = 30$$

Since $$x = \frac{1100 - 20y}{25}$$

on substitution $$x = \frac{1100 - 600}{25} = \frac{500}{25}$$

$\therefore$ $$x = 20.$$

$\therefore$ the total number of books sold was $30 + 20 = 50$, and the number originally sold at 25p was 20.

Exercise 22

1 There are two numbers, x and y, such that the sum of 2x and y is 34, while the sum of x and 2y is 32. What are the numbers?
2 There are two numbers such that if to 3 times the first, twice

the other is added, the sum is 72. Also if from 5 times the first number, 3 times the other is subtracted, the result is 44. What are the numbers?

3   One number is greater by 6 than twice another number, but 3 times the smaller number exceeds the greater by one. Find the numbers.

4   If from twice the greater of two numbers 17 is subtracted, the result is half the other number. If from half the greater number 1 is subtracted, the result is two thirds of the smaller number. What are the numbers?

5   In the equation $y = mx + b$, when $x = 3$, $y = 3$ and when $x = 5$, $y = 7$. Find the values of m and b and write down the equation. Then find y when $x = 6$.

6   Two quantities P and Q are connected by the formula:

$$P = \frac{m}{Q} + b$$

When $Q = 5$, $P = 14$ and when $Q = 2$, $P = 20$. Find m and b.

7   The force (E) applied to a machine and the resistance (R) to be overcome are connected by the law

$$E = a + bR.$$

It is found that when $E = 3.5$, $R = 5$ and when $E = 5.3$, $R = 8$. Find a and b. Then find E when $R = 8$.

8   It is known that $y = ax^2 + bx^3$; when $x = 2$, $y = 5.6$, and when $x = 3$, $y = 25$. Find the values of a and b.

9   The perimeter of a rectangular lawn is 32 m. It is reduced in size so that the length is four-fifths and the breadth is three-fourths of the original dimensions. The perimeter is then 25 m. What were the original length and breadth?

10   The bill for the telephone for a quarter can be expressed in the form

$$C = a + \frac{nb}{100}$$

where C is the total cost in pounds, a is a fixed charge, n the number of calls and b the price of each call in pence. When the number of calls was 104, the bill came to £5.83, and when the number was 67 the bill was £5.09. Find the fixed charge and the cost of each call.

11 The cost of 4 ties and 6 pairs of socks was £3.40, while that of 5 ties and 8 pairs of socks was £4.37. What were the prices of a tie and a pair of socks respectively?

12 The formula $S = ut + \frac{1}{2}at^2$ gives the distance (S) passed over by a moving body in t s.

In 4 s the body moves 88 m.

In 6 s the body moves 168 m.

Find the values of u and a and then find how far the body moves in 5 s.

9

Linear Inequalities

66 The idea of an inequality

In many situations we need to use words like *is less than* or *at least*.
Usually these will be statements involving numbers. For example:

'I have less than twenty pounds in my bank account'
'My ladder needs to be at least 4 m to reach the window'

It is not possible to deduce from these statements exactly how
much I have in the bank or how long my ladder is. For this reason
statements like this are sometimes called inequations, though it is
more usual to describe them as inequalities. We use special
symbols to show when an inequality exists between two quantities.

Examples:

4 *is less than* 7	$4 < 7$
3 m *is more than* 290 cm	$3 \text{ m} > 290 \text{ cm}$
My bank balance (B) *is less than* £20	$B < £20$
The ladder (L) *is at least* 4 metres	$L \geqslant 4 \text{ m}$

Note: the symbol $\geqslant$ is used to show that the length of the ladder
is either *equal to* 4 m or is *more than* 4 m.
A similar symbol $\leqslant$ is used to show when one quantity is *less than*
or *equal to* another quantity.
   Each of the above statements can be rewritten using the symbol
reversed.

$7 > 4$ 7 *is more than* 4

290 cm < 3 m   290 cm *is less than* 3 m
£20 > B   £20 *is more than* my bank balance
4 m ≤ L   4 m *is less than or equal to* L

67 Representing inequalities

'My ladder (L) is at least 4 m, but must be shorter than 6 m if it is to fit into my garage.'

The above statement can be shown using the inequality symbols as:

L ≥ 4 and L < 6   or as the single statement   4≤L < 6

One of the simplest ways of showing the actual numbers which satisfy this inequality is to use a number line.

On the number line the filled in ● is used to show that 4 can be included. However, since we were told that the ladder must be less than 6 m, an empty circle O is used at 6 to show that 6 m is not one of the possible lengths. The line joining the two circles indicates that the ladder can be any length between 4 m and 6 m.

Example: Represent the inequality on a 0 to 10 number line.
(a) n < 3   (b) n ≥ 6   (c) 2 < n ≤ 5

(*a*) n < 3

(*b*) n ≥ 6

(*c*) $2 < n \le 5$

Example 2: What inequality is represented on the 0 to 10 number line?

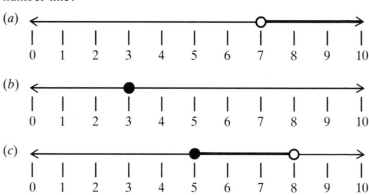

(*a*) represents $n > 7$, (*b*) represents $n \le 3$, (*c*) represents $5 \le 8$

Exercise 23

1 Write as an inequality:

    (*a*) 'My height (H) must be less than 190 cm to be a pilot.'
    (*b*) 'She must sell more than 50 pies (P) a week to make a profit.'
    (*c*) 'I will spend (S) at least £200 but less than £300.'
    (*d*) 'The trip (T) takes more than 3 hours but less than 5 hours.'

2 Show the inequality on a 0 to 10 number line.

    (*a*) $n > 7$, where n is a whole number.
    (*b*) $n \le 4$, where n is any number.
    (*c*) $3 \le n \le 8$, where n is a whole number.
    (*d*) $3 \le n \le 8$, where n is any number.

68 Solving inequalities

Many problems can be expressed in the form of an inequality statement.

'Tickets for a concert cost £4, £5, £6, £7 or £8. I want to buy two tickets and I have £20 of which £7 is needed for the train fares.'

Assuming the two tickets are to cost the same this is a situation which can be represented by the inequality, $2C + 7 \leq 20$.

Clearly this is a problem which can be solved by simply checking the cost of each type of ticket. I could in fact buy two £4, or two £5 or two £6, but not two £7 tickets, since $2 \times 6 + 7 < 20$ whereas $2 \times 7 + 7 > 20$. However with more complex situations we need to have a more general technique. This is very similar to the one used for solving equations.

$$2C + 7 \leq 20$$

subtracting 7 from each side gives

$$2C \leq 13$$

dividing each side by 2 gives

$$C \leq 6.5$$

so C can be 4 or 5 or 6 but not 7.

An inequality can be thought of as a balance where one side is heavier or lighter than the other. You can add or subtract the same number from each side without changing this situation. In the same way you could double or halve the numbers on each side, again without changing things.

However problems arise when you are dealing with negative numbers. It is helpful to look at a number line whch shows both positive and negative numbers to illustrate this difficulty.

The number line shows that the left-hand one of a pair of numbers on the line is smaller than the right-hand one of the pair. $3 < 5$, $-5 < 3$ and $-5 < -3$. It is important to realise that if $A > B$ then $-A < -B$.

In the first four examples below the inequality sign does not change.

$4 < 6$ adding 2 to each side gives $4 + 2 < 6 + 2$
                                                        i.e. $6 < 8$ $\checkmark$
$4 < 6$ subtracting 2 from each side gives $4 - 2 < 6 - 2$
                                                        i.e. $2 < 4$ $\checkmark$
$4 < 6$ doubling each side gives $4 \times 2 < 6 \times 2$
                                                        i.e. $8 < 12$ $\checkmark$
$4 < 6$ halving each side gives $4 \div 2 < 6 \div 2$
                                                        i.e. $2 < 3$ $\checkmark$

however if we were to multiply each side of $4 < 6$ by -2 we would get $4 \times (-2) = -8$ on the left side and $6 \times (-2) = -12$ on the right side. So in this case we need to change the inequality sign, since $-8 > -12$.

You can manipulate inequalities in the same way as equations, except you need to remember that when you multiply or divide each side of an inequality by *a negative number* you must then change the sign of the inequality.

Example 1: Solve the inequality $3x - 5 < 2x + 8$

First add 5 to each side $3x < 2x + 13$
Now subtract 2x from each side $x < 13$
So the solution is those values of x which are *less than* 13.

Example 2: Solve the inequality $3x + 5 > 5x - 9$

Method 1

subtract 5 from each side $3x > 5x - 14$
subtract 5x from each side $-2x > -14$
divide each side by -2
  and change the sign $x < 7$

So the solution is those values of x which are *less than* 7.

Note: in *Method 1* it is necessary to change the sign of the inequality since we are dividing by a negative number.

Method 2

add 9 to each side $3x + 14 > 5x$
subtract 3x from each side $14 > 2x$
divide each side by 2 $7 > x$
rewrite the inequality $x < 7$

So the solution is those values of x which are *less than* 7.

In *Method 2* the strategy was to collect the x terms on the side where they would be positive. Provided you are happy rewriting the inequality $7 > x$ as $x < 7$ this method is preferable and less likely to lead to mistakes.

Example 3 Solve the inequality $2(3x + 5) + 1 \geq 4x - 9$

First remove the brackets $\qquad 6x + 10 + 1 \geq 4x - 9$
Now subtract 11 from each side $\qquad\qquad 6x \geq 4x - 20$
Now subtract 4x from each side $\qquad\qquad 2x \geq -20$
Finally divide each side by 2 $\qquad\qquad\quad x \geq -10$

So the solution is those values of x which are *greater than* or *equal to* -10.

Exercise 24

1 Solve the inequality
(a) $2x - 5 > 9$ $\qquad\qquad$ (b) $3x + 7 < 16$

2 Solve the inequality
(a) $5 > x - 3$ $\qquad\qquad$ (b) $7 - x \geq 4$

3 Solve the inequality
(a) $2x < x + 8$ $\qquad\qquad$ (b) $5 + 2x \geq 3x$

4 Solve the inequality
(a) $3x + 2 < x + 8$ $\qquad\qquad$ (b) $2x + 7 \leq 5x + 1$

5 Solve the inequality
(a) $7 - 2x > 3x - 8$ $\qquad\qquad$ (b) $4 - 3x > 19 - 2x$

6 Solve the inequality $2(3x + 4) - 17 > 15$
7 Solve the inequality $3(x + 7) + 2 \leq 7(x - 2) + 5$
8 Solve the inequality $5(3x - 2) + 6 < 8(x - 7) + 3x - 12$

Inequalities involving fractions can be treated in a similar way, but again it is very important to be careful and for you to be aware of when you might be multiplying or dividing by a negative number. The next example shows one of the possible hidden traps that sometimes occur.

Example 4: Solve the inequality $\dfrac{8}{x} < 2$

This inequality asks us to find the values of x which when divided into 8 give a result which is less than 2. Clearly x can be any number which is greater than 4. For example 8 ÷ 5 is 1.6 which is less than 2. However x could also be any negative number, since 8 divided by a negative number will also be a negative number and all negative numbers are less than 2.

If we were simply to multiply both sides of the inequality by x, this would give 8 < 2x or x > 4 as the solution. This however is only partly correct. Remember we can only multiply by x without changing the inequality sign if x is a positive number. If x were negative, which is the other part of the solution, then the inequality would become 8 > 2x which is certainly true.

69 Simultaneous inequalities

In Chapter 8 you were introduced to the idea of a linear equation which involved two variables, and also how to find the solution for a pair of such simultaneous linear equations. We shall look at the corresponding work with inequalities involving two variables later in this book. However we should now consider the case where we might need to satisfy both of two simple linear inequalities at the same time.

Example 1: Find the values of x which satisfy both these inequalities

$$3x + 2 > 8 \text{ and } 5x - 3 < 27$$

Solving the first we get: 3x > 6 or x > 3
Solving the second we get: 5x < 30 or x < 6

So x must be a number which is *both greater than* 3 and *also less than* 6. So x can be any number *between* 3 and 6 but not including either 3 or 6.

This can also be shown as the single statement 3 < x < 6, or by using *square* brackets as]3, 6[. The outwards facing brackets indicate that the 3 and 6 are not included in the solution.

Using a number line the required solution is:

Example 2: Find the positive values of x which satisfy both these inequalities.

$$3(x + 5) > 8x \text{ and } 5(x - 3) < 27 - x$$

Solving the first we get: $3x + 15 \geqslant 8x$ or $15 \geqslant 5x$ or $x \leqslant 3$
Solving the second we get: $5x - 15 < 27 - x$ or $6x < 42$
or $x < 7$

So x must be a positive number which is *less than or equal to* 3 and which is also *less than* 7. The only possible vlaues of x are those which are *greater than or equal to* 0 abd *less than or equal to* 3.

So $0 \leqslant x \leqslant 3$. This can also be shown using square brackets as [0, 3]. The inwards facing brackets indicate that the 0 and 3 are included.

Using a number line the required solution is:

Note: $0 < x \leqslant 3$ would be shown as]0, 3] with one bracket of each type.

Exercise 25

1 Write the inequality using square brackets:

    (*a*) $2 < x < 7$ (*b*) $2 \leqslant x \leqslant 7$
    (*c*) $2 \leqslant x < 7$ (*d*) $2 < x \leqslant 7$

2 Rewrite using inequality signs:

    (*a*)]3, 6[(*b*) [5, 8]   (*c*)]−2, 5]   (*d*) [−2, 5[

3 Find the positive values of x which satisfy both the inequalities:

$$2x < x + 8 \text{ and } 5 + 2x \geqslant 3x$$

4 Find the positive values of x which satisfy both the inequalities:

$$3x - 2 < x + 8 \text{ and } 2x + 7 \leqslant 5x + 1$$

5 Find the positive values of x which satisfy both the inequalities:

$$3(x + 5) < 8x \text{ and } 5(x - 3) < 27 - x$$

With more complex situations it may be necessary to use a number line to show which numbers satisfy both inequalities.

Example 3: Find the values of x which satisfy both inequalities.

$$2 < x < 7 \text{ and } 4 < x < 9$$

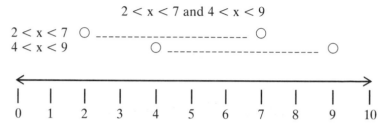

The numbers which satisfy both inequalities at the same time are the numbers between 4 and 7, i.e. $4 < x < 7$ or $]4, 7[$.

Example 4: Find the values of x which satisfy both inequalities.

$$-2 < x \leqslant 3 \text{ and } -1 < x \leqslant 5$$

The numbers which satisfy both inequalities at the same time are the numbers between -1 and 3, including 3 but excluding -1, i.e., $-1 < x \leqslant 3$.

Exercise 26

1 Find the values of x which satisfy both inequalities:

   (a) $4 < x < 8$, $3 < x < 7$ (b) $-2 < x < 1$, $-3 < x < 7$
   (c) $[-3, 8]$ and $]4, 12]$

2 Solve the simultaneous inequalities: $2(3x - 4) < x + 7$ and $-3 \leqslant x < 5$.

10

Graphical Representation
of Quantities

70 The object of graphical work

The graphical or pictorial representation of statistics and the comparison of magnitude by means of various graphical devices are familiar features of modern life. Rows and columns of figures or groups of large or very small numbers are not always readily grasped.

Accordingly, the use of lines or columns or other figures drawn to scale which appeal to the brain through the eye, is found to be an effective method of enabling many people to realise not only the quantities themselves but the deductions to be drawn from them.

The graphical method, developed on scientific and mathematical lines, is also largely employed in mathematics and science to illustrate certain important underlying principles. The following examples, arranged progressively, will help you to understand the various forms of graphical representation.

71 The column graph

The first example, Fig. 11, is a reproduction of an actual graphical advertisement, issued by the Cement Makers' Federation to show the variations in the deliveries of cement over a period of 11 years. It is an example of what is called a column graph.

Fig. 11.

Along the straight line OX equal spaces are marked off which indicate, in succession, the eleven years from 1929 to 1939 inclusive.

In the spaces so formed are constructed a series of columns of the same width and whose height, on the scale chosen, represents the number of tons of cement manufactured in each year. In order that these heights can be easily read a scale is marked along the OY, perpendicular to OX, in which each unit represents a million tons.

This assemblage of columns conveys very effectively, in a general way, the rise and fall of the production of cement during the eleven years. It also makes it possible to ascertain at a glance the approximate amount of cement produced in any one of the years.

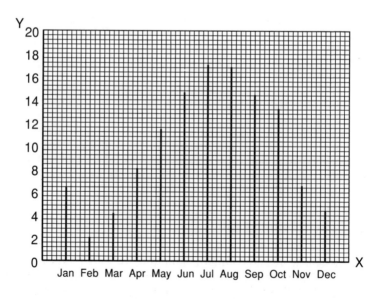

Fig. 12.

72 A straight-line graph

The next method illustrated is more commonly used than that of the column graph. The following table shows the mean temperature for each month throughout the year.

(1969) Month	Jan.	Feb.	Mar.	Apr.	May	June	July	Aug.	Sept.	Oct.	Nov.	Dec.
Mean temp. in °C	6.2	2.0	4.1	7.9	11.6	14.3	17.0	16.6	14.1	13.3	6.2	4.2

These are exhibited graphically in Fig. 12, and the method is as follows. Draw a straight line OX, and at suitable equal intervals mark points corresponding to the twelve months.

Draw OY at right angles to OX, and with a suitable scale mark points corresponding to temperatures from 0° to 20°.

To save time and secure accuracy, specially ruled paper, called 'graph paper', is employed.

At each of the 12 points on OX draw lines perpendicular to OY to represent the corresponding temperature as indicated on the scale on OY.

These straight lines take the place of the columns in Fig. 11.

This diagram shows at a glance not only the mean temperature for each month, but gives a clear picture of the rise and fall of average temperature throughout the year, the highest and the lowest.

73 A graph

It will be seen that when graph paper is used it is not necessary to draw the lines perpendicular to OX (as was done in Fig. 12). A point will mark the top of such a line, if it were drawn.

We now join these points by straight lines as shown in Fig. 13. The succession of lines thus formed is called a graph.

This is much more useful than the series of perpendicular lines in Fig. 12, and lends itself to important developments, as will be seen in later examples. Not only does it show more vividly the rise or fall each month, the highest and lowest temperatures, etc., but other features are apparent. For example, we get an idea of the average rate at which temperature is rising or falling during each month. This is indicated by the slope of the line corresponding to the month. Between November and December the slope of the line is very slight, the drop in temperature being small. Between March and April the steep slope shows a rapid increase in temperature. The month of the sharpest rise is April, when the slope is greatest. The slope is reversed when the temperature falls, and the month when the fall in temperature is greatest is evidently January. This is rather odd and January seems to have been an exceptionally warm month in 1969; the oddness of the temperature is reflected in the oddness of the graph, which kinks at this point.

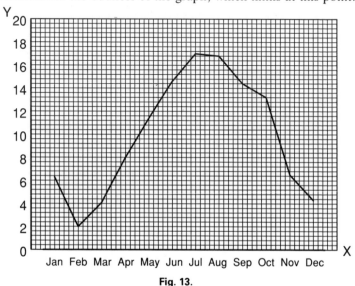

Fig. 13.

74 Examples of graphs and their uses

Graphs are employed in almost every branch of knowledge, and a full treatment of them is impossible in a volume of this size. The following typical elementary graphs may serve to illustrate their nature and their uses. The first example is from Insurance.

Example 1: The annual premiums charged by an Insurance Company for a policy of £100, at various ages, are as follows:

Age:	25	30	35	40	45	50
Premium in £s	2.33	2.59	2.91	3.31	3.81	4.53

The amounts of the premiums have been taken to the nearest penny. The method of 'plotting' these values, as it is called, is as follows.

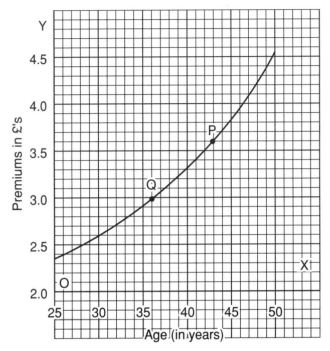

Fig. 14.

Two straight lines, OX, OY, are drawn at right angles (Fig. 14). Ages are marked off on OX, one small division representing a year. These are named as shown. Premiums are marked on OY, one small division representing 10p.

The lines OX and OY are called axes, OX being called the axis of x, and OY the axis of y, for reasons which will be apparent at later stages in the graphical work.

As there is no premium less than £2, we begin marking the premiums by placing the mark for £2 at O; similarly on the x axis we begin the ages with twenty-five years at O. By this means space on the paper is used to best advantage.

In succession we find the points which show the premium corresponding to the ages given. These are marked with dots. You should take squared paper and get these points for yourself. The process of obtaining the points is called 'plotting the points'.

When this is done it will be seen that the points appear to lie on a regular, smooth curve, and this has been drawn on the figure. An important point now arises.

If we are correct in assuming that points corresponding to the premiums for every fifth year lie on a regular smooth curve, then the values of these would seem to have been calculated according to a definite law or formula. If that is the case, then the premiums for the intervening years, which are not given in the table, would be calculated according to the same formula. It is therefore a reasonable deduction that the points corresponding to these, if plotted, would lie on the curve already obtained.

Accordingly, if we note the intersection of the curve with the perpendicular line corresponding to any intervening year, the position of this point with reference to the scale on the y axis indicates the amount of the premium for the year selected.

Thus, considering the point P, corresponding to the forty-third year, the premium as shown on the y axis is £3.60. Coversely, when the premium is £3 at Q, the corresponding age is thirty-six.

This process of obtaining values from the graph which lie between those plotted is called interpolation.

75   An example from Electricity

The table below shows the resistances in ohms for given lengths of wire of the same material and cross-section.

Length in metres	100	120	170	220
Resistance in ohms	2.5	3	4.25	5.5

Draw the graph which shows the relation between resistances and length and find the resistance for a length of 200 m.

Two straight lines OX and OY are drawn as before (Fig. 15).

Lengths are marked along OX, beginning with 100 m at O, and each small division representing 5 m.

Resistances are marked along OY, five divisions representing 1 ohm and

∴ each division representing 0.2 ohm.

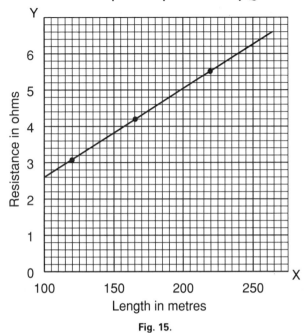

Fig. 15.

Using corresponding values from the table, points are plotted as before.

It will be seen that these points lie on a straight line.

As in the preceding example, we deduce that a definite law connects resistance and length.

Using the method of interpolation we see that when the length is 200 m the resistance is 5 ohms.

The graph being a straight line, it is easily produced as in the figure. This enables us to find the resistances corresponding to lengths beyond those given.

For example, when the length is 250 m the resistance is a little greater than 6.2 ohms.

This is called extrapolation.

76   An example from Mechanics

The following table shows the distances passed over in the times indicated by a body starting from rest. Draw a graph to show the relation between time and distance.

Time in seconds	0	1	2	3	4	5
Distance in metres	0	2	8	18	32	50

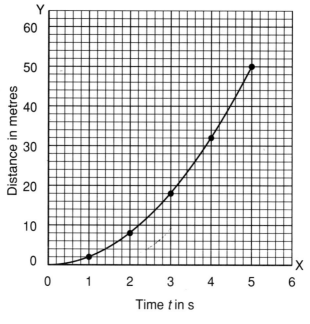

Fig. 16.

Drawing the axes OX and OY, units are selected in accordance with the range of numbers to be represented (see Fig. 16).

The points plotted are indicated by small dots. They appear to lie on a regular curve, and this is drawn to include all the points given in the table.

As in previous examples, the regularity of the curve suggests that time and distance are connected by a definite formula. This will be familiar to those students who have studied mechanics.

The curve is known as a parabola. Reasoning as in the previous example we may use the curve for interpolation. The following examples illustrate this.

Example 1: Find the distance passed over in 3.6 s.

From the curve this appears to be 26 m. The actual distance by formula is 25.92 m.

Example 2: How long will the body take to travel 42 m?

Finding the point which marks 42 m on the y axis, and then looking for the corresponding point on the curve, this is seen, from the scale on the x axis, to represent 4.6 s.

The Mechanics formula gives 4.58 s.

Exercise 27

1   The following table gives the values of exported manufactured goods of a certain type in years as specified.

Year:	1960	1961	1962	1963	1964	1965
Value in million £s	6.0	5.2	4.9	5.1	4.8	5.6

Show the variations by means of a graph.

2   The temperatures between 1000 h and 2000 h taken every two hours were as follows:

Time	1000	1200	1400	1600	1800	2000
Temp °C	8	9.1	11.9	13.7	12.1	9.0

Show these in a graph and find the probable temperature at 1300 h.

3   The expectation of life in years for males and females in this country at various ages is shown in the following table:

Age:	20	30	40	50	60	70
Expectation (Males)	47.2	38.5	29.8	21.8	14.7	8.8
Expectation (Females)	50.3	41.6	32.8	24.4	16.7	10.2

From your graph estimate the expectation of life for males and females at (1) 35 years, (2) 55 years.

4 The average mass of boys of different ages is given in the following tables. Draw a graph to illustrate them.

Age (years):	11	12	13	14	15
Mass (kg)	40	42.5	46	50.5	57

From the graph we find:

(1) The probable average mass of a boy of $12\frac{1}{2}$ years.
(2) The probable average mass at 16 years.
(3) If a boy of $13\frac{1}{2}$ years has mass 46.5 kg, about how much is he below average?

5 The populations of Salford CB and Poole MB at certain times is shown in the table below. Draw, using the same axes and scales, graphs showing the change in the two populations.

(1) From your graphs, when are the populations likely to be equal?
(2) What will this equal population be?

Draw the graph showing the difference in the two populations.

(3) When does this show the difference as being zero?

[*N.B.* In order to use the same axes and scales as before, draw the graph of 'difference + 100 000': the 'zero-line' will then be 100 000.]

Population in thousands

Year	1951	1961	1966	1968	1969
Salford	177	154	144	140	138
Poole	83	92	97	99	102

6 A train starts from rest, and its speed at intervals during the first minute is given in the following table:

Time (s)	0	5	10	20	30	40	50	60
Speed (m/s)	0	8.5	14.6	23	29.2	33.6	37	39

Draw a graph showing the relation between time and speed. Does it appear that a definite law connects these?

What is your estimate for the speed after 23 s?

7 The following table gives the annual premium £P payable during life by a man aged A years next birthday for a whole life assurance of £1000. Draw the graph.

A	20	25	30	35	40	45	50	55
P	11.60	13.10	13.30	17.50	21	25.50	31.50	40

A mistake was made in working out these figures: find from your graph which entry is wrong and estimate the correct entry. Estimate the annual premium for men aged 24, 42, 53 next birthday.

8 The following table shows the sales of gas for public lighting in the North West in the years 1963/4–1968/9. What is the likely sale in 1969/70?

Gas sales in kilotherms

Year	63/4	64/5	65/6	66/7	67/8	68/9
Gas	1933	1466	1017	723	552	434

9 If $y = \frac{1}{2}(x - 3)(x + 1)$, work out values of y for successive whole number values of x from $x = -3$ to $x = 5$. Draw a graph of $y = \frac{1}{2}(x - 3)(x + 1)$. From your graph, read off the two values of x that make y equal to 1.

11

The Law of a Straight Line; Co-ordinates

77   The straight-line graph

Among the graphs considered in the previous chapter were some which were straight lines. This regular arrangement of the plotted points suggested that a definite law governed the relations between the quantities, corresponding values of which were the basis of the plotting. This is an important principle which calls for further investigation.

The following example will serve as a starting-point.

The proprietor of a restaurant calculated the following figures showing the connection between his net profits and the number of customers. Exhibit the connection by a graph.

Number of Customers	240	270	300	350	380
Net profit in £s	0.50	2.00	3.50	6.00	7.50

Note: 'Net profit' means total receipts from customers, less expenses.

Let x = the number of customers.

Let y = the net profit.

Choosing two axes as usual, values of x will be plotted on the x axis, values of y on the y axis.

When plotted as in Fig. 17, the points are found to lie on a straight line.

Interpolation and extrapolation can be used as previously, but two important questions arise:

(1) What is the number of customers when there is no profit?

Zero on the profit (y) scale is shown where the graph cuts the x axis. Producing the graph it cuts at the point (A), where x = 230, i.e.,

when   x = 230, y = 0.

(2) What happens when the number of customers is less than 230?

If there is no profit when the number of cutomers is 230, there will be a loss when the number is below 230.

Let the straight line be produced below the x axis, as in Fig. 17. The amount of the loss, as with the profit, will be shown on the y axis, which must be produced.

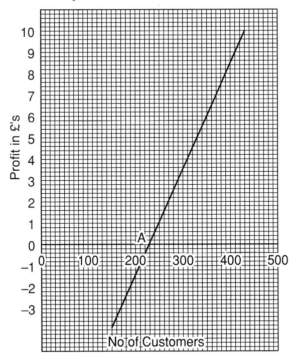

Fig. 17.

The question of indicating loss as contrasted with profit now arises as it did in Chapter 4. We must proceed in the same way. The loss will be marked with negative numbers on the scale, as shown. It will then be seen that on the y axis there is thus constructed part of a complete number scale (see Section 38).

From this scale, it appears that when x (the number of customers) is 180, y (the loss) is £2.50.

Problems will also arise when it will be necessary to use negative values of x, on the x axis produced. Consequently for a complete graph we shall require complete number scales on both axes, with zero common to both.

In the particular example above, a negative number of customers would have no intelligible meaning.

78   The law represented by a straight-line graph

Our next step is to discover the nature of the law which a straight-line graph represents and how it can be formulated. We will use the above problem as our example. In that problem it was stated that the net profit was equal to the total receipts less the expenses. But the total receipts equal (the number of customers) × (the average amount paid by each).

Let £a = the average amount thus paid.

Then since

  x = the number of customers,

 ax = the total amount paid in pounds.

Let £b = the expenses.

Then net profits are (ax − b) pounds, i.e.

$$y = ax - b.$$

This gives the value of y in terms of a, x, and b, and is the form of the law connecting them.

Of the four letters, a and b remain unchanged, while the number of customers (x) and consequently the profit (y) vary.

∴ the law is not completely stated until we know the values of a and b.

The method of doing this is suggested by example 2, Section 65.

Two pairs of corresponding values of x and y can be obtained from the table of values or from the graph.

For example, when

$$x = 320, y = 4.50$$

when $\qquad x = 250, y = 1.00.$

Substituting these in

$$y = ax - b$$
$$4.5 = 320a - b \qquad (1)$$
$$1 = 250 - b \qquad (2)$$

Subtracting (2) from (1)

$$3.5 = 70a$$

and $\qquad a = 0.05.$

Substituting in (1) we get

$$b = 11.5.$$

∴ the equation is $\qquad y = 0.05x - 11.5.$

This is the law of the straight line.

If it is correct it must be satisfied by any corresponding pair of values of x and y, and you should test this out.

In particular if $y = 0$, i.e. there is no profit

$$0.05x - 11.5 = 0$$

whence $\qquad 0.05x = 11.5$

and $\qquad x = 230.$

This agrees with the result found above.

Thus it can be demonstrated that the equation

$$y = 0.05x - 11.5$$

is satisfied by the coordinates of any point on the line and so is called the equation to the straight line.

79 Graph of an equation of the first degree

The equation which is represented by the straight line in the above problem (viz., $y = 0.05x - 11.5$) is of the first degree, that is, it contains no higher powers than the first of x and y.

Two questions now suggest themselves:

(1) Can every equation of the first degree in x and y be represented graphically by a straight line?

(2) Conversely, can every straight line be represented algebraically by an equation of the first degree?

The answers to these questions will be apparent later, but for the present we will confine ourselves to drawing the graphs of some typical equations of the first degree. From these the answer to the first question may be deduced.

Examples are given to illustrate the methods to be employed. In all of these, since both positive and negative values of x and y will be involved, complete number scales will be used on both axes.

Before calculating corresponding values of x and y, you should revise Section 62 if you are not familiar with the procedure.

80   Worked examples

Example 1: Draw the graph of the equation:

$$2y - 4x = 3.$$

This is not in the form used above, but it can readily be transformed thus:

$$2y - 4x = 3$$
$$2y = 4x + 3.$$
$$\therefore \qquad y = 2x + 1.5.$$

Giving suitable values to x, the corresponding values of y can be calculated, and so we get the following table:

x	-2	-1	0	1	1.8	2
y	-2.5	-0.5	1.5	3.5	5.1	5.5

A straight line is fixed by two points, but in drawing it from its equation three points should always be taken to check accuracy. In this example, as we are verifying the truth of a proposition, a number of points are taken, so as to make it clear that all such points lie on the straight line. The graph appears as shown in Fig. 18.

This could be checked by finding for any point on the line the corresponding values of x and y. These values should satisfy the equation.

Intercepts.

It should be noted that when

$$(1) \ x = 0, y = 1.5,$$

when   (2) y = 0, x = −0.75.

(1) 1.5 is called the intercept on the y axis (x = 0).
(2) −0.75 is called the intercept on the x axis (y = 0).

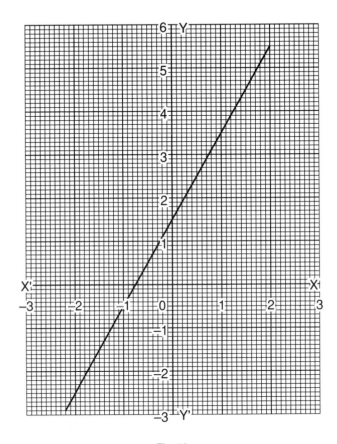

Fig. 18.

Example 2: Draw the graph of the equation:

$$2x + y = 1.$$

Transforming the equation we get:

$$y = −2x + 1 \qquad\qquad (A)$$

A table of corresponding values can be constructed as follows:

x	−2	−1	0	1	2	3
y	5	3	1	−1	−3	−5

With these values the straight line shown in Fig. 19 is drawn.

For intercepts on the axes

when x = 0, y = 1 (intercept on y axis)
when y = 0, x = 0.5 (intercept on x axis).

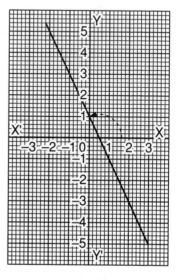

Fig. 19.

It should be noted from this example that when the coefficient of x in the equation arranged as in (A) is negative, the angle which the straight line makes with the x axis in an anti-clockwise direction is greater than a right angle. When the coefficient is positive, as in Fig. 18, this angle is less than a right angle.

Exercise 28

Draw the graphs which are represented by the following equations. In each case find the intercepts on (1) the y axis, (2) the x axis.

1 $y = x + 2$
2 $y = 1.5x - 1$
3 $4y = 6x - 5$
4 $3x + 2y = 6$
5 $\frac{x}{3} + \frac{y}{2} = 2$
6 $2(x - 3) = 4(y - 1)$

7 The equation $y = ax + b$ is satisfied by the following pairs of values of x and y:

   (1) x = 1, y = 5
   (2) x = 2, y = 7

Find the values of a and b, and substitute in the equation.

Draw the graph of this equation and find its intercept on the y axis.

8   A straight line makes an intercept of 2 on the y axis and of 4 on the x axis. Find its equation.

81   Position in a plane; co-ordinates

When a point is 'plotted', as in graphs previously considered, its position on the graph was fixed by the corresponding values of x and y which were given in the table. For example, let P be a point such that in the table x = 2, y = 3. When plotted, the point appears as in Fig. 20, where:

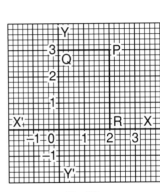

PQ is 2 units in length and parallel to OX,

and

PR is 3 units in length and parallel to OY.

The intersection of these two straight lines fixes the position not only in the graph, but also relative to the axes OX and OY. The position of any other point can be similarly determined when its distances from the two axes are known.

Fig. 20.

By such means might a man discover where, in a field, he had hidden a treasure. By remembering the distances OR and OQ along two boundaries of the field, he can 'plot the point P' in the field. A little reflection will convince you that the boundaries need not be at right angles to one another.

The distances PQ and PR (Fig. 20) which thus fix the position of a point are called the coordinates of P with respect to the two axes.

PQ, parallel to the x axis, is called the x co-ordinate (or the abscissa).

PR, parallel to the y axis, is called the y co-ordinate (or the ordinate).

The notation employed to denote co-ordinates is (2, 3) or in general (x, y). The x co-ordinate is always placed first inside the bracket.

Thus (5, 2) would represent the point in which x = 5, y = 2. In this way the positions of points relative to two axes can be described concisely. If, however, all points are to be included, complete number scales must be used on both axes, as shown in Fig. 21. In this way four divisions, called quadrants, are found in the plane; numbered I, II, III, IV as shown.

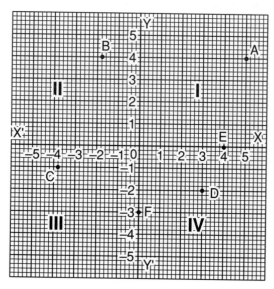

Fig. 21.

The signs of the co-ordinates are regulated by the positive and negative parts of the number scales.

The rule of signs is as follows:

If (x, y) be the co-ordinates of a point:

> x is +ve when measured to the right of O.
> x is −ve when measured to the left of O.
> y is +ve when measured up from O.
> y is −ve when measured down from O.

As examples the co-ordinates of A, B, C, D in the figure are:

A is (5, 4), B is (−2, 4), C is (−4, −1), D is (3, −2).

Points on the axes.

If a point is on the x axis the y co-ordinate is zero. Thus E is (4, 0).

If a point is on the y axis the x co-ordinate is zero. Thus F is (0, −3).

The intersection of the axes, O, is called the origin. Since it lies on both axes, its co-ordinates are (0, 0).

Thus the position of every point on a plane relative to two axes can be determined by co-ordinates.

Latitude and longitude are a practical example of the use of co-ordinates. They describe the position of a place with reference to the equator and the meridian through Greenwich as axes.

The introduction of co-ordinates was due to Descartes, who published his book on Analytical Geometry in 1638.

Exercise 29

1 Write down the co-ordinates of the points in Fig. 22 marked A, B, C . . . G.

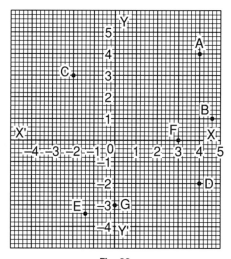

Fig. 22.

2 Join OA and FC in Fig. 22, and find the co-ordinates of their point of intersection.

3 Plot the points (3, 1), (1, 3), (0, 3), (0, −3), (4, 2.6), (−2, 1), (−4, −2), (3, 0), (−2, 0).

4   Draw the straight lines joining the points (3, 1) and (1, 3) and also (−2, 1) and (4, 2.6) as plotted in the previous question. What are the co-ordinates of the point of intersection of the two lines?

5   Plot the points (−3, 2), (0, 2), (2, 2), (4, 2). What do you notice about these points?

6   Plot the points (3, 3), (1, 1), (−1, −1), (−2, −2). What do you notice about these points?

7   Draw a straight line through (3, 0) parallel to YOY′. What do you notice about the co-ordinates of points on this line?

82   A straight line as a locus

Let A (Fig. 23) be a point such that its co-ordinates are equal. Let $(x_1 y_1)$ be its co-ordinates.

Then $y_1 = x_1.$

Join OA and draw AP perpendicular to OX.

Then $AP = OP.$

∴ by geometry $\angle OAP = \angle AOP$

since ∟APO is a right angle.

∴ each of the angles OAP, AOP is 45°.

∴ A lies on a straight line passing through the origin and making 45° with the axis OX.

Let B be any other point with equal co-ordinates (x_2, y_2) so that $BQ = OQ$, i.e., $y_2 = x_2.$

Then, for the same reasons as above, B also lies on a straight line passing through the origin and making 45° with the x axis.

This must be the straight line OA, since only one straight line can pass through the origin and make an angle of 45° with the x axis.

Similarly all other points with equal co-ordinates lie on the same straight line – i.e., the straight line OA, produced, is the locus of all points with equal co-ordinates.

These co-ordinates all satisfy the equation $y = x$, which is therefore the equation of the line.

This equation can be written in the form

$$\frac{y}{x} = 1.$$

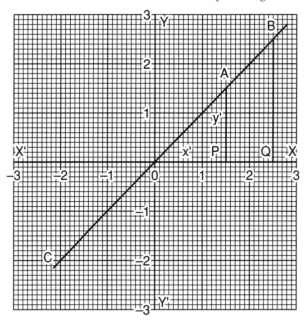

Fig. 23.

∴ for any point such as A,

$$\frac{AP}{OP} = 1.$$

This ratio is constant for all points on the straight line, and is called the gradient of the line.

A similar result holds for every straight line, consequently a straight line is a graph which has a constant gradient.

To include all points with equal co-ordinates the straight line must be produced into the opposite (3rd) quadrant.

Then for any point on this line the co-ordinates are equal, but they are both negative. Thus for the point C (Fig. 23) these co-ordinates are $(-2, -2)$ and the gradient is $\frac{-2}{-2}$, i.e., unity.

The student who has learnt sufficient geometry, or trigonometry, will know, without the above demonstration, that the ratio of $\frac{AP}{OP}$ is constant for any point A. See also Section 176.

83 Equation of any straight line passing through the origin

The conclusions reached above apply equally to all straight lines through the origin. The lines differ only in the gradient – i.e., in the value of $\frac{y}{x}$.

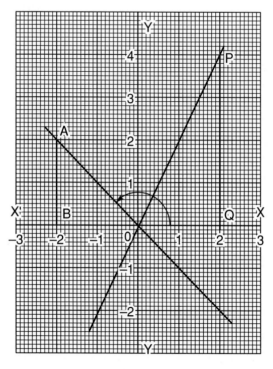

Fig. 24.

For example, if the gradient is 2, then $\frac{y}{x} = 2$ and $y = 2x$.

This is shown in Fig. 24. For any point P, on the line OP, is

the ratio $\qquad \frac{PQ}{OQ} = \frac{4}{2} = 2.$

Generally, if the gradient be denoted by m we have

$$\frac{y}{x} = m.$$

$$\therefore \qquad\qquad y = mx.$$

This is the general form of the equation of any straight line through the origin, where m denotes the gradient.

Negative gradient. If m is negative, the line will pass through the 2nd and 4th quadrants.

Thus if m = −1 the equation is

$$y = -x.$$

The straight line is shown in Fig. 24.

Considering any point A, the gradient is

$$\frac{AB}{OB} = \frac{+2}{-2} = -1.$$

It should be noted that the angle made with the axis of x is 135°, angles being always measured in an anti-clockwise direction.

84 Graphs of straight lines not passing through the origin

In Fig. 25 the straight line AOB is the graph of y = x.

If we now plot the graph of y = x + 2, it is evident that for any value of x the value of y in y = x + 2 is greater by 2 than the corresponding value of y in y = x.

∴ the line for y = x + 2 must be parallel to y = x, but each point is two units higher in the y scale. Thus in Fig. 25 the point A is raised to A′, the origin to D, B to B′, etc.

A′DB′ therefore represents the graph of y = x + 2.

The straight line y = x + 2 is the locus of all points whose co-ordinates are such that the y co-ordinate = the x co-ordinate + 2. It has the same gradient as y = x, but its intercept on the axis of y is +2.

Similarly the line y = x − 3 is parallel to the line y = x, with each point on it lowered by 3 units in the y scale.

Evidently then we can generalise and state that y = x + b will always represent a straight line parallel to y = x and with an intercept of b units on the y axis for any value of b.

The same conclusions hold for lines with different gradients. For example, the equation y = 2x + b will always represent a straight line parallel to y = 2x, i.e., having the same gradient, and with

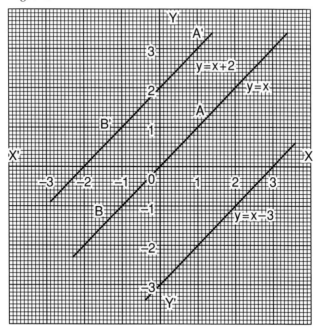

Fig. 25.

an intercept of b units on the y axis.

Generalising, let m = the gradient of a straight line.

Then y = mx + b always represents a straight line parallel to y = mx, i.e., with the same gradient, and with an intercept of b units on the y axis.

Examples:

(1) y = 4x − 7 is a straight line whose gradient is 4, and whose intercept on the y axis is −7.
(2) y = 0.05x − 11.5 (see Section 78) represents a straight line of gradient 0.05 and intercept on the y axis equal to −11.5.

As shown in Section 80, every equation of the first degree in two unknowns can be reduced to the form y = mx + b.

We therefore conclude that the graph of every equation of the first degree in two unknowns is a straight line.

Further, it is evident that the equation is satisfied by the co-ordinates of any point on the straight line.

85   Graphical solution of simultaneous equations

The conclusions reached in the previous paragraph can be used, as shown in the following example, to solve simultaneous equations of the first degree.

Example: Solve the equations:

$$x + 2y = 5 \tag{1}$$
$$3x - 2y = 7 \tag{2}$$

(1) Draw the graph of $x + 2y = 5$. The table of co-ordinates is as follows:

x	0	− 1	2	5
y	2.5	3	1.5	0

Note: When $x = 0$, y is the intercept on the x axis; when $y = 0$, x is the intercept on the x axis. It is useful to obtain these two points.

The graph is the straight line marked A in Fig. 26.

(2) Draw the graph of $3x - 2y = 7$. The table of values is:

x	0	− 1	$\frac{7}{8}$	4
y	− 3.5	− 2	0	2.5

The graph is the straight line marked B.

Applying the conclusions, reached in Section 84:

Line A contains all those points whose co-ordinates satisfy the equation $x + 2y = 5$.

Line B contains all those points whose co-ordinates satisfy the equation $3x - 2y = 7$.

There is one point, and one point only, whose co-ordinates satisfy both equations.

That point is P, the intersection of the two graphs.

The co-ordinates of P, by inspection, are (3, 1).

∴ the solution of the equation is

$$x = 3, y = 1.$$

You should compare the above conclusions with the algebraical treatment in Section 61.

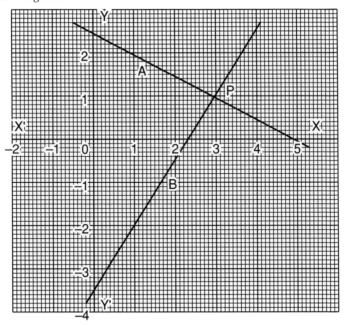

Fig. 26.

Exercise 30

1 With the same axes draw the graphs of:

    (1) $y = x$ (2) $y = 2x$ (3) $y = \frac{1}{2}x$
    (4) $y = -x$ (5) $y = -2x$

2 With the same axes draw the graphs of:

    (1) $y = x$ (2) $y = x + 1$ (3) $y = x + 3$
    (4) $y = x - 1$

3 With the same axes draw the graphs of:

    (1) $y = \frac{1}{2}x$ (2) $y = \frac{1}{2}x + 2$ (3) $y = \frac{1}{2}x - 1$

4 With the same axes draw the graphs of:

    (1) $y = x + 2$ (2) $y = 2x + 2$
    (3) $y = -x + 2$ (4) $y = \frac{1}{2}x + 2$

5 Draw the graphs of:

    (1) $2x + y = 3$ (2) $x - 2y = 4$
    (3) $5x + 2y = 10$ (4) $4x - 5y = 10$

    In each case find the intercepts on the x axis and on the y axis.

6 Solve graphically the following pairs of equations and check by algebraical solutions:

    (1) $x + y = 7$ (2) $2x + y = 7$
        $x - y = 1$ $3x - 5y = 4$
    (3) $2x + 3y = 2$
        $4x + y = -6$

7 The straight line whose equation is $y = ax - 1$ passes through the point (2, 5). What is the value of a? What is the intercept on the y axis?

8 The straight line whose equation is $y = 2x + b$ passes through the point (1, 3). What is the value of b? Draw the straight line. What is its intercept on the y axis?

9 The points (1, 1) and (2, 4) lie on the straight line whose equation is $y = ax + b$. Find a and b and write down the equation. What is the intercept of the straight line on the y axis?

    (*Hint*: See example 2 of Section 65.)

10 Draw the straight line $y = 2x + 1$.

    Draw the straight line parallel to it and passing through the point (0, 3). What is the equation of this line?

11 Find the equation of the straight line passing through the points (1, 3) and $\left(2, \dfrac{11}{3}\right)$. What is the gradient of the line?

12 Find the equation of the straight line passing through the points (2, 4) and (3, 5). Find its gradient.

12

Using Inequalities to Define Regions

86 Defining regions

On the grid below the line x = 3 and the line y = 4 have been marked.

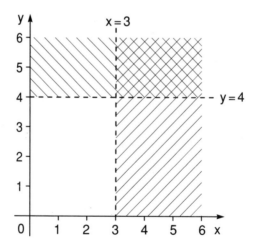

All the points whose x co-ordinate is greater than 3, shaded $///$, lie to the right of the line x = 3, whilst all the points whose co-ordinate is less than 3 lie to the left of the line x = 3.

Similarly all the points whose y co-ordinate is greater than 4, shaded \\\ , lie above the line y = 4, whilst all the points whose y co-ordinate is less than 4 lie below the line y = 4.

The region which is shaded ✕✕✕ is defined by the inequalities x > 3 and y > 4.

The region which is unshaded is defined by the inequalities x < 3 and y < 4.

The other two regions on this grid can be defined in a similar way.

87   Regions above and below straight lines

In Chapter 11 we saw how any equation of the form y = mx + c could be represented by a straight line with a gradient of m and an intercept on the y-axis of c.

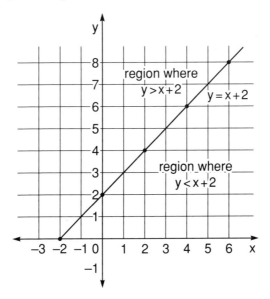

This straight line, if extended indefinitely in each direction, divides the page into two regions, one above the line and the other below the line.

The points in the region above the line y = x + 2 are described by the inequality y > x + 2, whilst the points in the region below the line are described by y < x + 2. The points on the line itself satisfy y = x + 2.

Example 1: Show on a graph the region described by the inequalities y > x and y < 2x − 3.

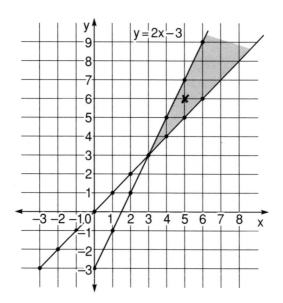

The points in the region y < 2x − 3 lie *below* the line y = 2x − 3.
The points in the region y > x lie *above* the line y = x.
The required region is shaded in the top right-hand part of the diagram.

Note the point (5, 6) lies in this region. Check: 6 > 5 and 6 < 2 × 5 − 3.

Example 2: Show on a graph and find one point in the region described by the inequalities y > x − 2, y < 2x + 1 and x + y < 6.

The points in the region y > x − 2 lie *above* the line y = x − 2.
The points in the region y < 2x + 1 lie *below* the line y = 2x + 1.
The points in the region x + y < 6 lie *below* the line x + y = 6.
The required region is shaded in the middle of the diagram.
One point in this region is (3, 2). Check: 2 > 3 − 2, 2 < 2 × 3 + 1 and 3 + 2 < 6.

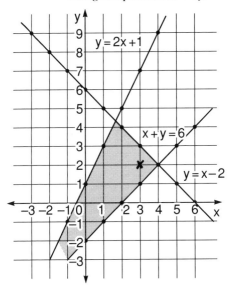

Exercise 31

1 Draw the given line and then shade the required region.
In each case find whether the point (3, 1) is in the region or not.

(*a*) line $x + y = 7$, region $x + y < 7$
(*b*) line $y = 2x + 1$, region $y > 2x + 1$
(*c*) line $y = 3x - 7$, region $y < 3x - 7$
(*d*) line $2x + 3 = 12$, region $2x + 3y > 12$

2 Draw the given line and then shade the required region.

(*a*) line $y = 2x$, region $y > 2x$
(*b*) line $3x + 2y = 24$, region $3x + 2y < 24$

Shade also the region where both $y > 2x$ and $3x + 2y < 24$.
Find whether the points (2, 5) and (6, 1) are in this region or not.

3 Draw the given line and then shade the required region.

(*a*) line $y = x + 23$, region $y < x + 2$
(*b*) line $y = 3x - 4$, region $y < 3x - 4$

Shade also the region where both $y < x + 2$ and $y < 3x - 4$.
Find whether the points (3, 4) and (4, 6) are in this region or not.

4 Draw the lines x + 2y = 12 and 2x + y = 10.
Mark the region where:

    (*a*) x + 2y > 12 and 2x + y < 10
    (*b*) x + 2y < 12 and 2x + y > 10
    (*c*) x + 2y < 12 and 2x + y < 10
    (*d*) x + 2y > 12 and 2x + y > 10

5 Draw the lines y = x + 3, y = 3x − 1 and x + y = 11.
Shade the region where y > x + 3, y < 3x − 1 and x + y < 11.
Find whether the points (3, 6) and (3, 7) are in this region or not.

6 Using the diagram below say which region is described by:

    (*a*) E   (*b*) D   (*c*) C

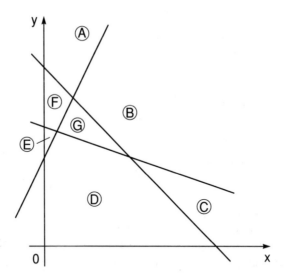

7 Draw the lines x + y = 7, y = 2x + 5 and x + 4y = 12.
Mark the region where:

    (*a*) x + y ⩽ 7, y ⩽ 2x + 5 and x + 4y ⩾ 12
    (*b*) x + y ⩾ 7, y ⩽ 2x + 5 and x + 4y ⩾ 12
    (*c*) x + y ⩽ 7, y ⩾ 2x + 5 and x + 4y ⩾ 12

8 Shade on a graph the region where:

(*a*) x + y < 10, 2x + y > 12 and x + 4y ≥ 16
(*b*) x + y > 10, 2x + y > 12 and x + 4y ≥ 16
(*c*) x + y < 10, 2x + y < 12 and x + 4y ≤ 16

88 Greatest or least values in a region

Example: Shade on a graph the region where x + 3y ≤ 24 and
3x + y < 21. Find at what points in this region the expression
x + y takes on its largest value, with x and y being whole numbers.

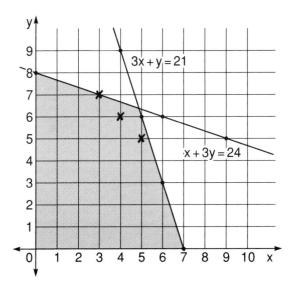

The points in the shaded region for which x + y is largest are
(3, 7), (4, 6) and (5, 5). Points on the line 3x + y = 21 are not
included since we want 3x + y < 21 and not 3x + y ≤ 21.

Note that the three points where x + y is largest lie on the line
x + y = 10.

Exercise 32

1 Look again at the previous graph.
Identify the region where 3x + y > 21 and x + 3y > 24.
At which whole number point(s) is x + y smallest in this
region?

2 Shade on a graph the region where $2x + y \leqslant 10$ and $x + 2y < 12$.

Find at what point(s) in this region the expression $x + y$ takes on its largest value, with x and y being whole numbers.

3 Use your graph from Question 2 to identify the region where $2x + y > 10$ and $x + 2y > 12$.

Find at what point(s) in this region the expression $x + y$ takes on its smallest value, with x and y being whole numbers.

4 Shade on a graph the region where $y \geqslant 2x - 10$ and $x + 3y < 18$.

Find at what point(s) in this region, with whole number values for x and y, the expression $x + y$ takes on its largest value.

Explain why the point (6, 4) is not included in the solution.

5 Use your graph from Question 4 to identify the region where $y < 2x - 10$ and $x + 3y \geqslant 18$.

Find at what point(s) in this region the expression $x + y$ takes on its smallest value, with x and y being whole numbers.

Explain why the point (7, 4) is not included in the solution.

89 Linear programming

The ideas in Section 87 can be applied to a range of practical problems which occur quite frequently in business and industry. The constraints on a situation are described by a set of inequalities and then it is required, for example, to maximise the profit or supply, or to minimise the labour required, within these constraints.

Example 1: A manufacturer makes two types of smoke detector. For model A, the materials cost £4, the labour £5 and the profit is £5. For model B, the materials cost £3, the labour £9 and the profit is £6. Altogether £240 is available for the materials and £450 is available for the labour. How many of each model should be made to give the maximum profit?

If a is the number of model A and b is the number of model B to be made then we can show the constraints on materials and labour as a the pair of inequalities: for materials $4a + 3b \leqslant 240$ and for labour $5a + 9b \leqslant 450$. We want to maximise the profit, which is given by £$(5a + 6b)$.

First we must show the two inequalities on a graph and find the region which corresponds to $4a + 3b \leqslant 240$ and $5a + 9b \leqslant 450$. In this example the region is the one below each of the two lines.

We now want to find the point in this region where $5a + 6b$ is greatest. It would be very tedious to check the value of $5a + 6b$ at each point but fortunately we can use our knowledge of parallel lines to help us. The lines $5a + 6b = 100$, $5a + 6b = 150$, $5a + 6b = 300$ etc. are all parallel. The second two are shown on the graph and you should note that as the profit gets larger the corresponding profit line moves up to the right. We want to find a point on the profit line which is furthest to the right in the required region. In this case the point is $(39, 28)$ and the profit £363.

Example 2: A farmer makes a composite feed from two different feeds so that the vitamin content of the final feed is at least:

Vitamin A – 8 units, Vitamin B – 6 units, Vitamin C – 3 units.

The vitamin content of each feed per kilogram is:

	Vitamin A	Vitamin B	Vitamin C
Feed 1	1	3	1
Feed 2	4	1	1

If the costs of each feed per kilogram are 120 pence and 90 pence respectively, how should the farmer make his feed to keep his costs to a minimum?

If we let the number of kilograms in each feed be x and y respectively, then we have the three constraints described by the inequalities:

$$\text{A: } x + 4y \geqslant 8, \text{ B: } 3x + y \geqslant 6 \text{ and C: } x + y \geqslant 3$$

The cost of making the composite feed, which we want to be as small as possible, is (120x + 90y) pence.

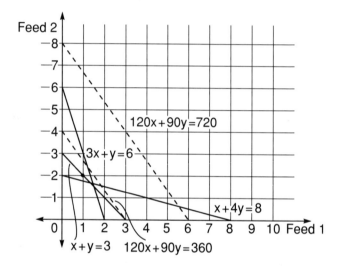

On the above diagram the two cost lines for 360p and 720p are shown as dotted lines. These correspond to 120x + 90y = 360 (i.e. 4x + 3y = 12) and 120x + 90y = 720 (i.e. 4x + 3y = 24). These lines are parallel and move down to the left as the cost decreases. The minimum cost line will be the last line which contains at least one point in the required region.

In this example the minimum cost occurs when the composite feed consists of 1 part of Feed 1 to 3 parts of Feed 2.

Exercise 33

1   For a camp of 144 children, two types of tent are available for hire. The large tent sleeps 8 and costs £32 per week, and the

small tent sleeps 3 and costs £8 per week.

The total number of tents must not exceed 36.

Using L for the number of large tents and S for the number of small tents write down inequalities to describe the constraints of:

   (*a*) the number of children   (*b*) the number of tents.

Write down an expression to shows the cost of hiring these tents.

If the cost of hiring is to be kept as small as possible, show the information on a graph and find the number of each type of tent that must be hired.

What will be the total cost of hiring for one week?

2 A market gardener has 100 hectares of land available for planting lettuces and/or spring onions, and he is prepared to outlay at most £2700. The initial outlay on each hectare of lettuces is £18, whilst that on each hectare of spring onions is £45.

Show this information as a pair of inequalities and represent it on a graph.

If the profit on each hectare of lettuces is £40 and on each hectare of spring onions is £60, find how he should allocate his land to make the maximum profit.

What is the greatest profit that he could make if he was prepared to use 120 hectares?

How many hectares must he be prepared to allocate to make it worth growing only lettuces?

3 A firm wants to buy a number of new cars of two types.

Type A costs £8000 and £48 per week to run.

Type B costs £9600 and £30 per week to run.

The firm has £72 000 to spend and wishes to keep its running costs to £360 per week.

If the firm must buy at least 4 cars of type A and 2 cars of type B, find the various possibilities for the number of each type of car that the firm could buy.

4 A firm with 10 drivers uses vans and lorries to make its deliveries.

A van costs £36 per day to run and can carry 600 kg.

A lorry costs £90 per day to run and can carry 1100 kg.

The firm wants to keep its daily running costs down to £540, but at the same time wants to be able to deliver at least 6600 kg.

Write down inequalities to describe the constraints of:

(*a*) drivers (*b*) running costs (*c*) carrying capacity.

Show this information on a graph and find the various possibilities for the number of vans and lorries if the firm decides to use at least 5 vans.

5 Represent on one graph the set of points (x, y) which satisfy the inequalities: $x \geq 10, y \geq 2, x + 4y \geq 24$ and $2x + 3y \geq 36$. Show also the region where $x + 2y \leq 30$ as well.

Hence find the points in this region, with whole number values for x and y, at which $x + 2y$ takes its smallest value.

13

Multiplication of Algebraical Expressions

90   (1) When one factor consists of one term

This has been considered in Section 27, when it was shown that

$$x(a + b) = xa + xb.$$

91   (2) Product of binomial expressions – i.e., with two terms.

A typical example is $(x + a)(y + b)$.

As in Section 27, a geometrical illustration will help to make clear what is the product and how it is obtained.

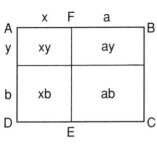

Fig. 27.

In Fig. 27 ABCD is a rectangle with the sides $(x + a)$ and $(y + b)$ units of length, and divided to represent x, a, y, and b units of length. Lines are drawn parallel to the sides dividing the whole rectangle into smaller ones whose areas represent the products xy, xb, ay and ab by means of their areas.

The area of the whole rectangle

$$= (x + a)(y + b) \text{ sq. units.}$$

Also the area of the whole rectangle

$$= \text{area of ADEF} + \text{area of EFBC}$$
$$= x(y + b) + a(y + b).$$

$$\therefore \quad (x + a)(y + b) = x(y + b) + a(y + b)$$
$$= xy + bx + ay + ab.$$

This grouping suggests the method of multiplying $(x + a)(y + b)$ algebraically.

The second binomial $(y + b)$ is multiplied in turn by each term of the first factor. The sum of these is the final product.

As the order in multiplication is immaterial so far as the final product is concerned, this could also have been obtained as follows. Writing the factors in the reverse order:

$$(y + b)(x + a) = y(x + a) + b(x + a)$$
$$= xy + ay + bx + ab.$$

This is illustrated in Fig. 27.

Worked examples

$$(a + b)(c + d) = a(c + d) + b(c + d)$$
$$= ac + ad + bc + bd.$$
$$(x + 5)(y + 3) = x(y + 3) + 5(y + 3)$$
$$= xy + 3x + 5y + 15.$$
$$(a + b)(c - d) = a(c - d) + b(c - d)$$
$$= ac - ad + bc - bd.$$
$$(a - x)(b - y) = a(b - y) - x(b - y)$$
$$= ab - ay - bx + xy.$$
$$(x - 4)(y + 3) = x(y + 3) - 4(y + 3)$$
$$= xy + 3x - 4y - 12.$$

If the first terms in each factor are alike, the same method is followed, the product being simplified afterwards if necessary. Thus:

$$(x + a)(x + b) = x(x + b) + a(x + b)$$
$$= x^2 + bx + ax + ab.$$

This could be expressed as

$$x^2 + (b + a)x + ab.$$

This result suggests a quick way of obtaining the product mentally. The coefficient of x in the answer is the sum of a and b. The last term is their product.

When a and b are numbers the sum and product will be evaluated, and the expression simplified.

Examples

$$(x + 6)(x + 5) = x(x + 5) + 6(x + 5)$$
$$= x^2 + 5x + 6x + 30$$
$$= x^2 + 11x + 30.$$
$$(a + 9)(a + 4) = a^2 + (9 + 4)a + 9 \times 4$$
$$= a^2 + 13a + 36.$$
$$(x + 2)(x - 7) = x(x - 7) + 2(x - 7)$$
$$= x^2 - 7x + 2x - 14$$
$$= x^2 + x(-7 + 2) - 14$$
$$= x^2 - 5x - 14.$$
$$(a - 8)(a - 3) = a(a - 3) - 8(a - 3)$$
$$= a^2 - 3a - 8a + 24$$
$$= a^2 - 11a + 24.$$
$$(x - 8)(x - 2) = x^2 + (-8 - 2)x + (-8) \times (-2)$$
$$= x^2 - 10x + 16.$$
$$(x - 8)(x + 2) = x^2 + (-8 + 2)x + (-8 \times 2)$$
$$= x^2 - 6x - 16.$$
$$(x + 8)(x - 2) = x^2 + (+8 - 2)x + \{+8 \times (-2)\}$$
$$= x^2 + 6x - 16.$$

92   When the coefficients of the first terms are not unity

The rule still holds. Thus:

$$(px + a)(qx + b) = px(qx + b) + a(qx + b)$$
$$= pqx^2 + pbx + aqx + ab,$$

which may be written as

$$pqx^2 + (pb + aq)x + ab.$$

The form of the coefficients of x in the last line should be noted. It will be used later.

Numerical examples of this form are common. The following illustrations show how to apply the rule quickly.

Worked examples

Example 1: $(2x + 5)(3x + 4)$

$$= 2x(3x + 4) + 5(3x + 4)$$
$$= 6x^2 + x(2 \times 4 + 5 \times 3) + 20$$
$$= 6x^2 + 23x + 20.$$

The coefficient of the middle term can be obtained by multiplying as shown by the arrows below; then add the results:

$$(2x + 5)(3x + 4).$$

Example 2:

$$(3x + 7)(2x + 1).$$

Product $= 6x^2 + x\{(3 \times 1) + (7 \times 2)\} + 7$
$\qquad = 6x^2 + 17x + 7.$

Example 3:

$$(7x - 5)(2x + 3).$$

Product $= 14x^2 + x\{(7 \times 3) - (5 \times 2)\} - 15$
$\qquad = 14x^2 + 11x - 15.$

Example 4:

$$(3x - 2)(4x - 7).$$

Product $= 12x^2 + x\{(3 \times -7) + (-2 \times 4)\} + 14$
$\qquad = 12x^2 - 29x + 14.$

If any difficulty is experienced the working should be set as shown in the second line of Example 1.

93 Multiplication of a trinomial

The method shown in Section 91 may be usefully adapted to certain cases in which one of the factors is a trinomial.

Example 1: $(x + 2)(x^2 - x + 1)$

$\qquad = x(x^2 - x + 1) + 2(x^2 - x + 1)$
$\qquad = x^3 - x^2 + x + 2x^2 - 2x + 2$
$\qquad = x^3 + x^2 - x + 2 \ \text{(on collecting like terms)}.$

Example 2: $(a + b)(a^2 - ab + b^2)$

$$= a(a^2 - ab + b^2) + b(a^2 - ab + b^2)$$
$$= a^3 - a^2b + ab^2 + a^2b - ab^2 + b^3$$
$$= a^3 + b^3 \text{ (on collecting like terms).}$$

Exercise 34

Write down the following products:

1	$(a + x)(b + y)$	**2**	$(c + d)(e + f)$
3	$(ax + b)(cy + d)$	**4**	$(a - x)(b - y)$
5	$(x - y)(a - b)$	**6**	$(a - x)(b + y)$
7	$(a + x)(b - y)$	**8**	$(a + 2)(b + 3)$ ·
9	$(a - 2)(b - 3)$	**10**	$(a - 2)(b + 3)$
11	$(a + 2)(b - 3)$	**12**	$(x + 7)(x + 5)$
13	$(ab + 6)(ab + 3)$	**14**	$(x + 10)(x + 3)$
15	$(x - 10)(x - 3)$	**16**	$(x + 10)(x - 3)$
17	$(x - 10)(x + 3)$	**18**	$(p + 8)(p - 12)$
19	$(x - 4y)(x - 8y)$	**20**	$(x - 4y)(x + 8y)$
21	$(x + 4y)(x - 8y)$	**22**	$(a + 2b)(2a + 5b)$
23	$(3x - 4y)(3x - 5y)$	**24**	$(4x + 1)(7x + 2)$
25	$(2x - 3)(3x - 1)$	**26**	$(3x + 1)(3x - 4)$
27	$(1 + 3y)(1 - 4y)$	**28**	$(6x + 1)(3x - 5)$
29	$(7x - 3y)(2x + 5y)$	**30**	$(3a - 7b)(6a - 5b)$

31 Simplify $\{(x + y)(a + b) - (ay + yb)\}$ and divide the result by x.

32 Simplify $\{(a + b)(a - c) + bc\} \div a$.

33 Find the value of $(2x - y)(x + y) - (2x + y)(x - y)$ when $x = 3, y = 2$.

34 Simplify $(1 - 2x)(1 + 3y) - (1 - 2y)(1 + 3x)$ and find its value when $x = 0.1, y = 0.2$.

35 Find the following products:

(*a*) $(x - y)(x^2 + xy + y^2)$
(*b*) $(a + 2)(a^2 - 2a + 4)$
(*c*) $(1 + x)(1 - x + x^2)$
(*d*) $(x + a)(x^2 + 2ax + a^2)$

94 Square of a binomial expression

In Section 91 it was shown that

$$(x + a)(x + b) = x^2 + x(a + b) + ab.$$

Since this is true whatever the values of the letters, let b = a.

Then $(x + a)(x + a) = x^2 + x(a + a) + a \times a$
$$= x^2 + 2xa + a^2.$$

$\therefore$ $(x + a)^2 = x^2 + 2ax + a^2.$

If a be replaced by $-a$ throughout, then

$$(x - a)^2 = x^2 - 2ax + a^2.$$

Geometrical illustration.

Fig. 28 shows an illustration of the above result by means of a square whose side is x + a.

It is a modification of Fig. 27 and will be readily understood without further explanation.

Try to draw a similar figure to illustrate

$$(x - a)^2 = x^2 - 2ax + a^2.$$

Examples

$(x + 1)^2 = x^2 + 2x + 1$
$(x - 1)^2 = x^2 - 2x + 1$
$(a + 9)^2 = a^2 + 18a + 81$
$(1 - 5xy)^2 = 1 - 10xy + 25x^2y^2$
$(2x + 7y)^2 = 4x^2 + 28xy + 49y^2$
$(3a - 10b)^2 = 9a^2 - 60ab + 100b^2$

Fig. 28.

95 Square of a trinomial

In the product

$$(x + c)^2 = x^2 + 2cx + c^2.$$

Since x may have any value, replace it with a + b.
Then

$(a + b + c)^2 = (a + b)^2 + 2c(a + b) + c^2$
$$= a^2 + 2ab + b^2 + 2ac + 2bc + c^2,$$

or, re-arranging,

$$(a + b + c)^2 = a^2 + b^2 + c^2 + 2ab + 2bc + 2ac.$$

This may be stated in words as follows:

The square of a trinomial is equal to
(sum of squares of each term) + (twice the product of the three terms taken two at a time in every possible way).

Examples

$$(a + b + 1)^2 = a^2 + b^2 + 1 + 2ab + 2a + 2b$$
$$(x - y + z)^2$$
$$= x^2 + (-y)^2 + z^2 + 2(x \times -y) + 2(xz) + 2(-y \times z)$$
$$= x^2 + y^2 + z^2 - 2xy + 2xz - 2yz$$
$$(a - b - c)^2 = a^2 + b^2 + c^2 - 2ab - 2ac + 2bc$$
$$(x + y - 5)^2 = x^2 + y^2 + 25 - 10x - 10y + 2xy$$

Care should be taken in applying the laws of signs.

96 Cube of a binomial

$(a + b)^3$ may be written as $(a + b)(a + b)(a + b)$
$$= (a + b)(a^2 + 2ab + b^2).$$

Multiplying these as shown in Section 93
$$(a + b)^3 = a(a^2 + 2ab + b^2) + b(a^2 + 2ab + b^2)$$
$$= a^3 + 2a^2b + ab^2 + a^2b + 2ab^2 + b^3.$$

$\therefore$

$(a + b)^3 = a^3 + 3a^2b + 3ab^2 + b^3$ (on collecting like terms).

Similarly

$$(a - b)^3 = a^3 - 3a^2b + 3ab^2 - b^3.$$

Examples

$$(x + 1)^3 = x^3 + 3x^2 + 3x + 1$$
$$(x - 1)^3 = x^3 - 3x^2 + 3x - 1$$
$$(1 + a)^3 = 1 + 3a + 3a^2 + a^3$$
$$(1 - a)^3 = 1 - 3a + 3a^2 - a^3$$
$$(2x + 3y)^3 = (2x)^3 + \{3(2x)^2 \times 3y\} + \{3(2x) \times (3y)^2\} + (3y)^3$$
$$= 8x^3 + 36x^2y + 54xy^2 + 27y^3$$
$$(x - 3y)^3 = x^3 - 9x^2y + 27xy^2 - 27y^3$$

Note: In the cube of $(x - y)$ the signs are alternately positive and negative.

Exercise 35

Write down the following squares in full:

1 $(x + 2)^2$	**2** $(x - 2)^2$
3 $(a + 3b)^2$	**4** $(a - 3b)^2$
5 $(2x + y)^2$	**6** $(x - 2y)^2$
7 $(ab + 10)^2$	**8** $(xy - 3)^2$
9 $(4x + 5y)^2$	**10** $(4x - 5y)^2$
11 $(5xy + 6)^2$	**12** $(1 - 10x^2)^2$
13 $(5x^2 + 3y^2)^2$	**14** $(3xy - 2y^2)^2$
15 $\left(x + \dfrac{1}{y}\right)^2$	**16** $\left(\dfrac{1}{x} - \dfrac{1}{y}\right)^2$
17 $\left(a + \dfrac{1}{3}\right)^2$	**18** $\left(\dfrac{1}{2} - \dfrac{3y}{4}\right)^2$
19 $\{(x + y) + 1\}^2$	**20** $\{1 - (x - 2y)\}^2$
21 $(a + b - c)^2$	**22** $(x - y + z)^2$
23 $(2x + 3y - 5z)^2$	**24** $(4a - 2b - 1)^2$

Write down the following cubes in full:

25 $(x + y)^3$	**26** $(x - y)^3$
27 $(a + 2)^3$	**28** $(a - 2)^3$
29 $(p + q)^3$	**30** $(p - q)^3$
31 $(2x + y)^3$	**32** $(x - 2y)^3$
33 $(3a - 1)^3$	**34** $(1 - 3b)^3$

35 Simplify the following: $(x + y)^2 - (x^2 + y^2)$.

36 Simplify:

(1) $(a + b)^2 - (a - b)^2$
(2) $(a - b)^2 - (a + b)^2$

37 Simplify $(x + 10)^2 - (x - 10)^2$.

38 If $x = 3y + 1$, express $x^2 + 4x + 4$ in terms of y and find its value when $y = 1$.

39 A square lawn the side of which is x m is surrounded by a path (Fig. 29) which is a m wide. Find an expression for the area of the path in terms of x and a. What is the area of the path when $x = 30$ m and $a = 2$ m?

40 If the lawn of the previous question were a rectangle, x m by y m and the path were a m wide, find an expression for its area.

Fig. 29.

97 Product of sum and difference

If a and b be any two numbers, then the product of their sum and difference is expressed by $(a + b)(a - b)$.

Using the method of Section 91 to find the value of this

$$(a + b)(a - b) = a(a - b) + b(a - b)$$
$$= a^2 - ab + ab - b^2$$

$$\therefore \qquad (a + b)(a - b) = a^2 - b^2$$

This important result can be expressed in words as follows:

The product of the sum and difference of two numbers is equal to the difference of their squares.

EXAM.

Examples

$$(x + 9)(x - 9) = x^2 - 81$$
$$(ab + 10)(ab - 10) = a^2b^2 - 100$$
$$(4x + 5)(4x - 5) = 16x^2 - 25$$
$$(1 + x)(1 - x) = 1 - x^2$$
$$(a - \tfrac{1}{2})(a + \tfrac{1}{2}) = a^2 - \tfrac{1}{4}$$
$$\left(\frac{a}{3} + \frac{b}{4}\right)\left(\frac{a}{3} - \frac{b}{4}\right) = \frac{a^2}{9} - \frac{b^2}{16}$$

$$\{(a + b) + c\}\{(a + b) - c\} = (a + b)^2 - c^2$$
$$\{a + (b + c)\}\{a - (b + c)\} = a^2 - (b + c)^2$$

Exercise 36

Write down the following products:

1 $(a + x)(a - x)$
2 $(p + q)(p - q)$
3 $(a + 2b)(a - 2b)$
4 $(4x + 3)(4x - 3)$
5 $(2x + 1)(2x - 1)$
6 $(1 + 6x)(1 - 6x)$
7 $(1 + a^2)(1 - a^2)$
8 $(2x^2 + 1)(2x^2 - 1)$
9 $(x^2 + y^2)(x^2 - y^2)$
10 $(3xy + 2)(3xy - 2)$
11 $(12xy + 1)(12xy - 1)$
12 $(\tfrac{1}{4}x - 7)(\tfrac{1}{4}x + 7)$
13 $\{(x + y) + z\}\{(x + y) - z\}$
14 $(a + x + y)(a + x - y)$
15 $(2a + 3b + 1)(2a + 3b - 1)$
16 $(x - 2y + 6)(x - 2y - 6)$
17 $\{a + 2(b + c)\}\{a - 2(b + c)\}$
18 $\{2x + 3(y + z)\}\{2x - 3(y + z)\}$
19 $(x + \tfrac{2}{3})(x - \tfrac{2}{3})$

20 $\left(\frac{x}{2} + \frac{y}{3}\right)\left(\frac{x}{2} - \frac{y}{3}\right)$

21 $(x + a + \tfrac{1}{2})(x + a - \tfrac{1}{2})$

14

Factors

98   The work of this chapter will be the converse of that in the previous one. That was concerned with methods of obtaining the products of certain algebraical expressions. In this chapter we shall seek to find the factors of expressions of different types.

A converse operation, in general, is more difficult than the direct one, and so it is in this case. By rules, on the whole simple and easily applied, the products of various kinds of factors are found. But in seeking to find the factors whose product produces a given expression, rules, even when they are formulated, are often long and tedious. In the main we rely on trial methods, which are, however, not haphazard but based on those rules by which the product was obtained.

It is always possible to obtain a product when the factors are given, but we cannot always find factors for expressions. Most expressions have no factors, and so we can deal only with special types, such as were obtained as products in the work of the preceding chapter.

99   Monomial factors – i.e., factors consisting of one term only

This is the converse of the theorem stated in Section 27. There we saw that

$$x(a + b) = xa + xb.$$

If we start with $xa + xb$ and wish to factorise it, we see by inspection that x is a factor of each term. It is therefore a factor of the whole expression. To find the other factor we divide each term by x and add the quotients. As a result we get a + b.

$$\therefore \qquad xa + xb = x(a + b).$$

In finding the factors by inspection we are guided by a knowledge of the process by which when (a + b) is multiplied by x the result is xa + xb.

100 Worked examples

Example 1: Find the factors of $6a^2 + 3ac$.

In this case we see that there is more than one factor common to each term.

(1) 3 is the highest common factor of the numerical coefficients.
(2) a is a factor of the other parts of each term.

$\therefore$ 3a is a factor of each term, and is therefore a factor of the whole expression. Dividing each term by it,

$$6a^2 + 3ac = 3a(2a + c).$$

Example 2: Factorise $5x^2y^2 - 10x^2y + 20y^2$.

The highest numerical factor common to each term is 5, and the only other factor is y.

$$\therefore \qquad 5x^2y^2 - 10x^2y + 20y^2 = 5y(x^2y - 2x^2 + 4y).$$

Exercise 37

Express the following as factors:

1	$6x + 12$	**2**	$3ab + 2a$
3	$4xy + 2y^2$	**4**	$6a^2 - 4ab$
5	$14x^2y^2 - 7xy$	**6**	$16 - 32a^2$
7	$a^2 - ab + ac$	**8**	$x^3 + 3x^2 - x$
9	$15a^3 - 5a^2b + 3a^2b^2$	**10**	$6a^2c - 15ac^2$
11	$a^2b + ab^2 - abc$	**12**	$\dfrac{bc^2}{6} - \dfrac{abc}{9}$

13 Fill in the blank in the following:

$$(7.4 \times 13^2) + (7.4 \times a^2) = 7.4(\qquad\qquad).$$

14 Calculate the following as easily as you can:

$$18.6^2 + 18.6 \times 1.4.$$

101 Binomial factors

In Section 91 we saw that

$$(x + a)(y + b) = x(y + b) + a(y + b)$$
$$= xy + bx + ay + ab.$$

If we require the factors of $xy + bx + ay + ab$ and similar expressions, we must work backwards through the steps shown above.

The first step is to reach the stage $x(y + b) + a(y + b)$.

To obtain this the four terms of the expression must be suitably arranged in two pairs such that:

(1) The terms in each pair have a common factor.
(2) When this common factor is taken out, the same expression must be left in each pair.

$\therefore$ we group $xy + bx + ay + ab$ in the form

$$(xy + bx) + (ay + ab)$$

Then taking out the common factor in each pair we get

$$x(y + b) + a(y + b).$$

Then $y + b$ is a factor of both parts of the expression and must therefore be a factor of the whole.

Thus we get, on taking it out,

$$(y + b)(x + a).$$

We will not apply this method to expressions of which the factors are not previously known, as they were in the example.

102 Worked examples

Example 1: Find the factors of $a^2 + cd + ad + ac$.

The first two terms have no common factor. Consequently the order of the terms could be changed in order to get two pairs, each with a common factor.

∴ we write

$$a^2 + cd + ad + ac = (a^2 + ad) + (cd + ac).$$

In this arrangement

(1) a is common to the terms of the first group and c to those of the second,
(2) the same expression a + d will be the other factor of each group.

∴ we get $\qquad a(a + d) + c(a + d)$

(a + d) being a factor of both parts is a factor of the whole.

∴ expression $\qquad = (a + d)(a + c).$

Note: Another possible arrangement was:

$$\begin{aligned}
&(a^2 + ac) + (ad + cd) \\
&= a(a + c) + d(a + c) \\
&= (a + c)(a + d).
\end{aligned}$$

There are always two possible ways of grouping.

Example 2: Factorise, if possible, ab + ac + bc + bd.

This example is given to show you where you might go astray.

The expression can be grouped as a(b + c) + b(c + d). But the expressions in the brackets in the two parts are different. It is a bad mistake, which is sometimes made, to write down the factors as (b + c)(c + d)(a + b). Not one of these is a factor of the expression. Moreover their product would be an expression of the third degree. The given expression is of the second degree.

On trying different groupings, it will be found that it is not possible to arrange in two groups, having the same factor in each group.

There are no factors of this expression.

Example 3: Find the factors of ab − 5a − 3b + 15.

By arrangement into suitable pairs:

$$\begin{aligned}
&ab - 5a - 3b + 15 \\
&= (ab - 5a) - (3b - 15) \\
&= a(b - 5) - 3(b - 5) \\
&= (b - 5)(a - 3).
\end{aligned}$$

It should be carefully noted that when an expression is placed in brackets with minus sign in front, as was done with $-3b + 15$ overleaf, the signs within the brackets must be changed. This is in agreement with the reverse rule given in Section 28, cases (3) and (4).

Exercise 38

Find the factors of:

1	$ax + ay + bx + by$	**2**	$pc + qc + pd + qd$
3	$ab - bd + ae - de$	**4**	$ax - cx - ay + cy$
5	$x^2 + px + qx + pq$	**6**	$x^2 - gx - hx + gh$
7	$ab + 5b + 6a + 30$	**8**	$ab - 5b - 6a + 30$
9	$ab - 5b + 6a - 30$	**10**	$2ab - 10a + 3b - 15$
11	$ax^2 + a^2x - ab - bx$	**12**	$x^2 + ax - bx - ab$

103   The form $x^2 + ax + b$

It was seen in Section 91 that the product of two factors such as $(x + 6)(x + 5)$ was $x^2 + 11x + 30$.

Reversing the process, we must now consider how to find the factors of $x^2 + 11x + 30$.

In the general case it was shown that

$$(x + a)(x + b) = x^2 + x(a + b) + ab.$$

In this product

(1) the coefficient of x is the sum of the numbers a and b,
(2) the term independent of x, i.e., ab, is the product of these numbers.

Consequently in finding the factors of an expression such as $x^2 + 11x + 30$, we must find by trial two factors of 30 whose sum is $+11$. They are $+5$ and $+6$.

∴ the factors are $(x + 5)(x + 6)$.

104   Worked examples

Example 1: Factorise $x^2 + 13x + 36$.

There are several pairs of factors of 36 – viz. (1×36), (2×18), (3×12), (4×9) and (6×6).

We look for the pair whose sum is $+13$; this pair is $(9, 4)$.

$\therefore$ $\qquad x^2 + 13x + 36 = (x + 4)(x + 9)$.

Example 2 Factorise $x^2 - 13x + 36$.

In this case the sign of the middle term is $-$ and of the last term is $+$.

$\therefore$ we must look for two negative factors of 36 whose sum is -13.

These are -9 and -4.

$\therefore$ $\qquad x^2 - 13x + 36 = (x - 9)(x - 4)$.

Example 3: Factorise $y^2 - 13y + 30$.

Proceeding as in the last example, we get:

$$y^2 - 13y + 30 = (y - 10)(y - 3).$$

Example 4: Factorise $x^2 - 5x - 36$.

When the last term is negative the factors of 36 must have opposite signs.

$\therefore$ the coefficient of x is the sum of a positive and negative number, and the negative number must be the greater numerically, or we might say that 5 is the difference of the two factors numerically.

The factors are clearly -9 and $+4$.

$\therefore$ $\qquad x^2 - 5x - 36 = (x - 9)(x + 4)$.

Note: The larger of the two numbers has the same sign as the middle term.

Example 5: Factorise $x^2 + 12x - 28$.

The two factors of 28 which differ by 12 are 14 and 2. They are of opposite signs, and their sum is $+12$.

$\therefore$ the factors required are $+14$ and -2.

$\therefore$ $\qquad x^2 + 12x - 28 = (x + 14)(x - 2)$.

Example 6: Factorise $a^2 - 8ab - 48b^2$.

The introduction of the second letter makes no difference to the method followed, but b will appear in the second term of each factor.

Thus $\qquad a^2 - 8ab - 48b^2 = (a - 12b)(a + 4b)$.

Exercise 39

Find the factors of:

1	$x^2 + 3x + 2$	**2**	$x^2 - 3x + 2$
3	$x^2 + 5x + 6$	**4**	$x^2 - 5x + 6$
5	$x^2 + 7x + 6$	**6**	$x^2 + 9x + 20$
7	$x^2 - 12xy + 20y^2$	**8**	$a^2 - 15ab + 36b^2$
9	$x^2y^2 + 15xy + 54$	**10**	$a^2b^2 - 19ab + 48$
11	$y^2 - 21y + 108$	**12**	$x^2 - 12xy + 35y^2$
13	$x^2 - x - 2$	**14**	$x^2 + x - 2$
15	$x^2 + xy - 6y^2$	**16**	$x^2 - xy - 6y^2$
17	$b^2 - 2b - 3$	**18**	$b^2 + 2b - 3$
19	$x^2 + 13x - 48$	**20**	$x^2 - 13x - 48$
21	$x^2 - xy - 110y^2$	**22**	$a^2 - 11a - 12$
23	$a^2 - a - 12$	**24**	$p^2 + p - 72$
25	$p^2 - 34p - 72$	**26**	$1 - 9x + 20x^2$
27	$1 - 8x - 20x^2$	**28**	$x^2y^2 - 3xy - 88$
29	$p^2 + 4p - 45$	**30**	$p^2 + pq - 56q^2$

105　The form $ax^2 + bx + c$

The factors of such an expression as this, where a, b, c are numbers, will be of the form which occurred in the converse operation in Section 92. These factors are best obtained by trial, as indicated in the following examples.

Example 1: Find the factors of $2x^2 + 7x + 3$.

Write down possible pairs of factors systematically and find the middle term as shown in Section 92 until the correct coefficient of x is found. In this example the possibilities are:

(1) $(2x + 3)(x + 1)$.

(2) $(2x + 1)(x + 3)$.

In (1) the coefficient of x in the product is 5.
In (2) the coefficient of x in the product is 7.

The second pair is therefore the correct one – i.e.,

$$2x^2 + 7x + 3 = (2x + 1)(x + 3).$$

Example 2: Find the factors of $6x^2 + 17x - 3$.

The minus sign of the last term indicates that the factors will have opposite signs in the second terms. This increases the number of possible pairs of factors. Among the possible pairs are:

(1) $(2x - 1)(3x + 3)$ or $(2x + 1)(3x - 3)$
(2) $(6x - 1)(x + 3)$ or $(6x + 1)(x - 3)$

Proceeding as shown in Section 92, we find that the first pair of (2) is the one required, since the coefficient of x is

$$(6 \times 3) + (1 \times -1) = 18 - 1 = +17.$$

$$\therefore \qquad 6x^2 + 17x - 3 = (6x - 1)(x + 3).$$

Example 3: Find the factors of $4x^2 - 17x - 15$.

The sign of the last term is again negative; therefore the factors will have opposite signs. Among the possibilities are:

(1) $(2x + 5)(x - 3)$ or $(2x - 5)(2x + 3)$
(2) $(4x + 5)(x - 3)$ or $(5x - 5)(x + 3)$
(3) $(4x + 3)(x - 5)$ or $(4x - 3)(x + 5)$

The first pair of (3) gives for the coefficient of x

$$(4 \times -5) + (3 \times 1) = -20 + 3 = -17.$$

$$\therefore \qquad 4x^2 - 17x - 15 = (4x + 3)(x - 5).$$

Exercise 40

Complete the following factors:

1 $3x^2 + 10x + 8 = (3x \quad)(x \quad)$
2 $12x^2 - 17x + 6 = (4x \quad)(3x \quad)$
3 $12x^2 - 28x - 5 = (6x \quad)(2x \quad)$
4 $9x^2 + 43x - 10 = (9x \quad)(x \quad)$

Find the factors of the following:

5 $2x^2 + 3x + 1$ **6** $3x^2 - 4x + 1$
7 $2x^2 + 5x + 2$ **8** $6x^2 + 5x + 1$
9 $4x^2 - 8x + 3$ **10** $5x^2 - 6x + 1$
11 $6x^2 - 11x + 3$ **12** $12x^2 + 11x + 2$
13 $2a^2 + a - 1$ **14** $2a^2 - a - 1$

15 $2a^2 - a - 6$ **16** $10b^2 - b - 2$
17 $10b^2 + b - 2$ **18** $8y^2 - 14y - 15$
19 $12x^2 + 5x - 2$ **20** $14c^2 - 17c - 6$

106   Squares of binomial expressions

A trinomial expression which is the square of a binomial can be recognised by applying the rule given in Section 94.

The standard forms are:

$$(a + b)^2 = a^2 + 2ab + b^2$$

and

$$(a - b)^2 = a^2 - 2ab + b^2.$$

If the trinomial is a square, the following conditions must be satisfied:

(1) The first and last terms, when the trinomial has been suitably arranged, are exact squares and positive.
(2) The middle term must be ± (twice the product of the square roots of the first and third terms).

Example 1: Write $x^2 + 6x + 9$ as a square of a binomial expression.

(1) The first and third terms are the squares of x and 3.
(2) The middle term is $+ (2 \times \sqrt{x^2} \times \sqrt{9}) = 6x$.

$\therefore$ $x^2 + 6x + 9 = (x + 3)^2.$

Similarly $x^2 - 6x + 9 = (x - 3)^2.$

Example 2: Is $4x^2 + 6x + 9$ a complete square?

(1) The first and third terms are the squares of 2x and 3.
(2) For a complete square the middle term should be
  $+ (2 \times \sqrt{4x^2} \times \sqrt{9}) = + 12x.$

But the middle term is $+ 6x$.
$\therefore$ the expression is not a complete square.

Example 3: Is $4x^2 - 20x + 25$ a complete square?

The middle term is

$$-2(\sqrt{4x^2}) \times \sqrt{25} = 2 \times 2x \times 5 = 20x.$$

$\therefore$ $4x^2 - 20x + 25 = (2x - 5)^2.$

107 Difference of two squares

In Section 97 it was shown that

$$(x + a)(x - a) = x^2 - a^2.$$

Conversely $\qquad (x^2 - a^2) = (x + a)(x - a).$

$\therefore$ an expression which is the difference of two squares has for its factors the sum and the difference of the numbers which are squared.

Note: No real factors can be found for the sum of two squares – i.e., $x^2 + a^2$.

108 Worked examples

Example 1: Factorise $100x^2 - 1$.

The numbers squared are $10x$ and 1.
$\therefore$ the factors are the sum and difference of these – i.e.,

$$100x^2 - 1 = (10x + 1)(10x - 1).$$

Example 2: Factorise $36a^2b^2 - 25$.

The numbers squared are $6ab$ and 5.

$\therefore \qquad 36a^2b^2 - 25 = (6ab + 5)(6ab - 5).$

Example 3: Factorise $(a + b)^2 - c^2$.

Although one of the terms squared is replaced by a binomial expression, the rule above still applies.

$$\therefore \qquad (a + b)^2 - c^2 = \{(a + b) + c\}\{(a + b) - c\}$$
$$= (a + b + c)(a + b - c).$$

Example 4: Factories $(a + b)^2 - (c - a)^2$.

$$(a + b)^2 - (c - a)^2 = \{(a + b) + (c - a)\}\{(a + b) - (c - a)\}$$
$$= (a + b + c - a)(a + b - c + a)$$
$$= (b + c)(2a + b - c).$$

109 Evaluation of formulae

The above rule is frequently employed in transforming formulae and in arithmetical calculations.

Example 1: Find the value of $47.5^2 - 22.5^2$.

$$47.5^2 - 22.5^2 = (47.5 + 22.5)(47.5 - 22.5)$$
$$= 70 \times 25$$
$$= 1750.$$

Example 2: Find the area of a ring between two concentric circles of radii 97 mm and 83 mm, respectively. The area of a circle is πr^2 where the radius is r.

The area of the ring is the difference between the areas of the two circles.

Difference in area

Difference in area $= \pi \times 97^2 - \pi \times 83^2$
$$= \pi(97^2 - 83^2) \text{ taking out common factor}$$
$$= \pi\{(97 + 83) \times (97 - 83)\}$$
$$= \pi \times 180 \times 14$$
$$= 2520\pi \text{ mm}^2.$$

Exercise 41

Write each of the following as the square of a binomial expression:

1 $p^2 + 2pq + q^2$ **2** $x^2 - 4xy + 4y^2$
3 $9x^2 + 6x + 1$ **4** $16x^2 - 40xy + 25y^2$
5 $x^2 + x + \frac{1}{4}$ **6** $\frac{a^2}{9} + \frac{ab}{3} + \frac{b^2}{4}$

Express as complete squares:

7 $(a + b)^2 + 4(a + b) + 4$
8 $(x - y)^2 - 10(x - y) + 25$

Find the factors of the following:

9 $x^2 - 100$ **10** $a^2b^2 - 25$
11 $4x^2 - 9y^2$ **12** $25a^2 - 16b^2$
13 $121x^2 - 36y^2$ **14** $144p^2 - 169q^2$
15 $25 - 16a^2$ **16** $1 - 225x^2$
17 $8a^2 - 50b^2$ **18** $3x^2 - 75$
19 $5x^2 - 45y^2$ **20** $(a + b)^2 - c^2$

21 $(x + 2y)^2 - 16z^2$ **22** $1 - \frac{81}{16}y^2$

23 $x^2 - (y + z)^2$ **24** $a^2 - (x - 2y)^2$
25 $(x - 8)^2 - 49$ **26** $(a + b)^2 - (a - b)^2$

Find the numerical values of the following:

27 $65^2 - 35^2$ **28** $82^2 - 68^2$

29 $49^2 - 39^2$ **30** $24^2 - 18^2$

31 $4.25^2 - 1.75^2$ **32** $17.5^2 - 12.5^2$

33 Find the value of $\pi(r_1^2 - r_2^2)$ when $\pi = \frac{22}{7}$, $r_1 = 12.5$, $r_2 = 8.5$.

34 If $S = \dfrac{v^2 - u^2}{2f}$ find S when $v = 14.5$, $u = 2.5$ and $f = 1.5$.

35 In the formula $w = k(D^2 - d^2)$ find w, when $k = 2.4$, $D = 8.5$, $d = 7.5$.

36 If $y = \dfrac{m(v^2 - u^2)}{2p}$ find y when $m = 24$, $v = 44$, $u = 16$ and $p = 32$.

37 If $s = \frac{1}{2}gt^2$ find the difference between the values of s when $t = 8$ and $t = 6$. Take $g = 10$.

38 In the right-angled triangle (Fig. 30) a and b represent the lengths of the sides containing the right angle and c the length of the hypotenuse. We know from geometry that

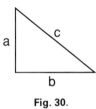

Fig. 30.

$$c^2 = a^2 + b^2.$$

(1) Find a when $c = 52$, $b = 48$.
(2) Find b when $c = 65$, $a = 25$.

110 Sum and differences of two cubes

In Section 93, Example 2, it was shown that

$$(a + b)(a^2 - ab + b^2) = a^3 + b^3.$$

By using the same method we get:

$$(a - b)(a^2 + ab + b^2) = a^3 - b^3.$$

These results enable us to obtain factors of the expressions:

$$a^3 + b^3 \text{ and } a^3 - b^3.$$

Re-arranging them we have:

$$a^3 + b^3 = (a + b)(a^2 - ab + b^2) \qquad \text{(A)}$$
$$a^3 - b^3 = (a - b)(a^2 + ab + b^2) \qquad \text{(B)}$$

It may aid the memory to note that

(1) In (A), the sum of two cubes, the first factor is the sum of the numbers.
(2) In (B), the difference of two cubes, the first factor is the difference of the numbers.

The other factors differ only in the sign of ab.

(1) When the first factor is the sum the sign is minus.
(2) When the first factor is the difference the sign is plus.

111   Worked examples

$$x^3 + 1 = (x + 1)(x^2 - x + 1)$$
$$x^3 - 1 = (x - 1)(x^2 + x + 1)$$
$$1 - x^3 = (1 - x)(1 + x + x^2)$$
$$a^3 - 27 = (a - 3)(a^2 + 3a + 9)$$
$$8x^3 + 125 = (2x + 5)(4x^2 - 10x + 25)$$

Exercise 42

Find the factors of:

1   $x^3 + c^3$ 2 $y^3 - a^3$
3   $1 + 8a^3$ 4 $x^3 - 64$
5   $8 + 27c^3$ 6 $R^3 - 1$
7   $m^3 - 125n^3$ 8 $x^3y^3 + \frac{1}{8}$
9   $\dfrac{1}{x^3} + \dfrac{1}{y^3}$ 10 $\dfrac{1}{x^3} - \dfrac{1}{y^3}$

15

Fractions

112 Algebraic fractions

In algebra fractions have the same fundamental meaning as in arithmetic, and are subject to the same rules in operations with them. They differ only in the use of letters in the algebraic forms.

An algebraic fraction is one in which the denominator is an algebraic expression. The numerator may or may not be algebraic in form.

Thus $\frac{2}{a}$ is an algebraic fraction, but $\frac{a}{2}$ is not. The latter means $\frac{1}{2}a$, and thus $\frac{1}{2}$ is merely a fractional coefficient.

Similarly $\frac{a}{a + b}$ is an algebraic fraction but $\frac{a + b}{6}$ is not.

113 Laws of fractions

It was stated above that algebraic fractions are subject to the same laws as arithmetical fractions. It is unnecessary to state these laws here, but a few illustrations may serve to remind the student how they operate. It was pointed out in Section 26 that

$$\frac{a}{b} = \frac{a \times m}{b \times m} \text{ and } \frac{a}{b} = \frac{a \div m}{b \div m} \, ,$$

where m is any number.

By means of this rule algebraic fractions can be simplified or reduced to lowest terms by dividing both numerator and denomi-

nator by the same number. This can be expressed by the phrase
'common factors are cancelled'. This was assumed in Section 22.

114 Reduction of fractions

Examples were given in Section 26, but further instances are now
given depending on the operations of the previous chapter.

Example 1: Simplify $\dfrac{a + b}{a^2 - b^2}$.

Using the rule of Section 107.

$$\frac{a + b}{a^2 - b^2} = \frac{a + b}{(a + b)(a - b)}$$
$$= \frac{1}{a - b} \text{ (on cancelling } (a + b)).$$

Example 2: Simplify $\dfrac{x^2 + 4x - 12}{x^2 + x - 6}$.

Factorising numerator and denominator,

$$\frac{x^2 + 4x - 12}{x^2 + x - 6} = \frac{(x - 2)(x + 6)}{(x - 2)(x + 3)}$$
$$= \frac{x + 6}{x + 3} .$$

The mistake is sometimes made of cancelling terms instead of
factors. Thus, there can be no cancelling with the 3 and 6 in the
above answer. Only factors of the whole expressions in numerator
and denominator can be cancelled.

Example 3: Simplify $\dfrac{3a(a^2 - 4ab + 4b^2)}{6a(a^2 + 3ab - 10b^2)}$.

Factorising, the fraction becomes

$$\frac{3a(a - 2b)^2}{6a(a + 5b)(a - 2b)} .$$

The factors $3a(a - 2b)$ are common to numerator and denominator. Cancelling them, the fraction becomes,

$$\frac{a - 2b}{2a(a + 5b)} .$$

Exercise 43

Simplify the following fractions:

1 $\dfrac{2xyz}{6x^2z^2}$

2 $\dfrac{15abd^3}{20a^3bd}$

3 $\dfrac{2x^2 + 2xy}{4xy - 4y^2}$

4 $\dfrac{3a^3 + 9ab}{6a^3b - 6a^2b^2}$

5 $\dfrac{6x^2y + 6xy^2}{4x^3y^2 + 4x^2y^3}$

6 $\dfrac{a^2 - b^2}{a^2 - ab}$

7 $\dfrac{2x - 6}{x^2 - 5x + 6}$

8 $\dfrac{x^2 - 9x + 20}{x^2 + x - 20}$

9 $\dfrac{a^3 + 6a^2b + 9ab^2}{a^2b + 5ab^2 + 6b^3}$

10 $\dfrac{x^3 - xy^2}{x^4 - xy^3}$

115   Multiplication and division

If necessary you should revise Section 26 and then proceed to the harder examples now given.

Example 1: Simplify $\dfrac{x}{x + 1} \div \dfrac{x^2}{x^2 - 1}$.

Factorising and inverting the second fraction:

$$\frac{x}{x + 1} \div \frac{x^2}{x^2 - 1} = \frac{x}{x + 1} \times \frac{(x + 1)(x - 1)}{x^2} .$$

Cancelling common factors we get:

$$\frac{x - 1}{x} .$$

Example 2: Simplify $\dfrac{x^4 - 27x}{x^2 - 9} \div \dfrac{x^2 + 3x + 9}{x + 3}$.

Factorising the expression is equal to:

$$\frac{x(x^3 - 27)}{x^2 - 9} \div \frac{x^2 + 3x + 9}{x + 3}$$
$$= \frac{x(x - 3)(x^2 + 3x + 9)}{(x + 3)(x - 3)} \times \frac{x + 3}{x^2 + 3x + 9}$$
$$= x \text{ (on cancelling factors).}$$

Exercise 44

Simplify the following fractions:

1 $\dfrac{xy}{x^2 - 1} \times \dfrac{x^2 - x}{y}$ **2** $xy \div \dfrac{x}{y + 1}$

3 $\dfrac{a^2 - b^2}{a^2 + 2ab + b^2} \times \dfrac{ab + b^2}{a^2 - ab}$ **4** $\dfrac{a^2 - 49}{a^2 - 9} \div \dfrac{a + 7}{a + 3}$

5 $\dfrac{2a^3 - 6a^2b}{2a^3 - 2ab^2} \times \dfrac{a^2 + 3ab + 2b^2}{a - 3b}$

6 $\dfrac{2x^2 + x - 1}{x - 2} \div \dfrac{2x^2 - x}{x^2 - 5x + 6}$

7 $\dfrac{2a^2 - 8}{a + 2} \div 2a - 4$

8 $\dfrac{6x^2 - x - 2}{8x + 4} \times \dfrac{2}{3x^2 - 5x + 2}$

9 Simplify $\dfrac{2b}{b^2 - 1} \times \dfrac{b^2 - 3b - 4}{b^2}$ and find its value when $b = 5$.

10 Simplify $\dfrac{x^2 + 6x + 9}{2x - 2} \times \dfrac{x^2 - 1}{x^2 + 4x + 3}$ and find its value when $x = 1.5$.

116　Addition and subtraction

The fundamental principles were examined in Section 25. We now proceed to more difficult examples.

Example 1: Simplify $\dfrac{a}{a - b} - \dfrac{a^2}{a^2 - b^2}$.

Factorising, the expression becomes

$$\frac{a}{a - b} - \frac{a^2}{(a + b)(a - b)} .$$

The least common denominator is $(a + b)(a - b)$.
$\therefore$ fraction

$$= \frac{a(a + b) - a^2}{(a + b)(a - b)}$$
$$= \frac{a^2 + ab - a^2}{(a + b)(a - b)}$$
$$= \frac{ab}{(a + b)(a - b)} .$$

Example 2: Simplify $\dfrac{3}{a - b} - \dfrac{2a + b}{a^2 - b^2}$.

Factorising, the expression becomes

$$\dfrac{3}{a - b} - \dfrac{2a + b}{(a + b)(a - b)}$$
$$= \dfrac{3(a + b) - (2a + b)}{(a + b)(a - b)} = \dfrac{3a + 3b - 2a - b}{(a + b)(a - b)}$$
$$= \dfrac{a + 2b}{(a + b)(a - b)}.$$

This example is intended to remind you of the note at the end of Section 30 relative to a minus sign before a fraction. The numerator, as $(2a + b)$ in the above example, should always be placed in a bracket on addition of the fraction, and the bracket removed afterwards.

Example 3: Simplify $\dfrac{x - 1}{x^2 - x - 2} - \dfrac{x + 2}{x^2 + 4x + 3}$.

Factorising, the expression becomes

$$\dfrac{x - 1}{(x - 2)(x + 1)} - \dfrac{x + 2}{(x + 3)(x + 1)}$$
$$= \dfrac{(x - 1)(x + 3) - (x + 2)(x - 2)}{(x + 1)(x - 2)(x + 3)}$$
$$= \dfrac{(x^2 + 2x - 3) - (x^2 - 4)}{(x + 1)(x - 2)(x + 3)}$$
$$= \dfrac{x^2 + 2x - 3 - x^2 + 4}{(x + 1)(x - 2)(x + 3)}$$
$$= \dfrac{2x + 1}{(x + 1)(x - 2)(x + 3)}.$$

Example 4: Simplify $\dfrac{x}{x - \dfrac{1}{x}}$.

$$\dfrac{x}{x - \dfrac{1}{x}} = \dfrac{x}{\dfrac{x^2 - 1}{x}} = x \div \dfrac{x^2 - 1}{x}$$
$$= x \times \dfrac{x}{x^2 - 1} = \dfrac{x^2}{x^2 - 1}.$$

Example 5: Convert the formula $\dfrac{1}{R} = \dfrac{1}{R_1} - \dfrac{1}{R_2}$ into one in which the subject is R.

$$\frac{1}{R} = \frac{1}{R_1} - \frac{1}{R_2}$$

i.e., $\quad \dfrac{1}{R} = \dfrac{R_2 - R_1}{R_1 R_2} .$

$\therefore \quad R = \dfrac{R_1 R_2}{R_2 - R_1} \left(\text{just as, if } \dfrac{1}{R} = \dfrac{1}{4}, \therefore R = 4 \right).$

Note: You should not make the mistake of inverting at the outset. The fractions on the right side must be added first.

Exercise 45

Simplify the following:

1 $\dfrac{1}{x} + \dfrac{1}{x - y}$

2 $\dfrac{2}{x - 1} - \dfrac{3}{x - 2}$

3 $\dfrac{2}{3(a - b)} + \dfrac{3}{2(a + b)}$

4 $1 - \dfrac{1}{x} + \dfrac{1}{x + 2}$

5 $3 - \dfrac{2}{x - 3} + \dfrac{1}{(x - 3)^2}$

6 $\dfrac{2x}{3(x + 1)} - \dfrac{3x}{2(x - 2)}$

7 $\dfrac{3}{1 - a} + \dfrac{4}{(1 - a)^2}$

8 $\dfrac{x + 3}{x - 3} - \dfrac{x - 3}{x + 3}$

9 $\dfrac{1}{x} - \dfrac{1}{x + 2y} - 1$

10 $\dfrac{b}{3a - b} - \dfrac{a}{3b} + 1$

11 $\dfrac{1}{1 - at} - \dfrac{1}{1 - bt}$

12 $\dfrac{1}{a + bt} - \dfrac{1}{x + yt}$

13 $\dfrac{1}{x + y} + \dfrac{1}{x - y} - \dfrac{1}{x^2 - y^2}$

14 $\dfrac{x}{x^2 - y^2} - \dfrac{y}{(x - y)^2} - \dfrac{1}{x + y}$

Transform the following formulae to those in which the subject is R:

15 $\dfrac{1}{R} = \dfrac{1}{p} + \dfrac{1}{q}$

16 $\dfrac{1}{R} = \dfrac{1}{p} - \dfrac{1}{q}$

17 $\dfrac{1}{R} = \dfrac{2}{R_1} + \dfrac{3}{R_2}$

18 $\dfrac{1}{R} = \dfrac{3}{R_1} - \dfrac{2}{R_2}$

19 If $\dfrac{1}{R} = \dfrac{1}{r + s} - \dfrac{1}{r - s}$ find R in terms of the other letters.

20 If $\dfrac{1}{P} = \dfrac{2}{p-q} + \dfrac{3}{p+q}$ find the value of 5P.

21 Simplify $P + Q - \dfrac{P^2 + Q^2}{P + Q}$.

22 Simplify $\dfrac{P^2 + Q^2}{P - Q} - P$.

23 If $\dfrac{1}{R} = \dfrac{1}{R_1} + \dfrac{1}{R_2} + \dfrac{1}{R_3}$ find R when $R_1 = 8.6$, $R_2 = 4.3$, and $R_3 = 3$.

24 Given $a(P - \tfrac{1}{2}Q) = b(Q - \tfrac{1}{2}P)$ rearrange the terms so as to express P in terms of the other quantities.

23 Given $I = \dfrac{nE}{R + nr}$ find n in terms of the other quantities. Find n when $I = 2$, $E = 1.8$, $R = 2.4$, $r = 0.5$.

117 Simple equations involving algebraical fractions

The following examples will illustrate the methods employed.

Example 1: Solve the equation:

$$\frac{3}{x-2} = \frac{5}{x-1}.$$

The least common denominator is $(x - 2)(x - 1)$.
We therefore multiply both sides by this.

Then $\qquad \dfrac{3(x - 2(x - 1))}{x - 2} = \dfrac{5(x - 2)(x - 1)}{x - 1}.$

Cancelling $\qquad 3(x - 1) = 5(x - 2)$

or $\qquad 3x - 3 = 5x - 10.$

$\therefore \qquad 2x = 7$

and $\qquad x = 3.5.$

Example 2: Solve for n the equation:

$$\frac{1}{n-2} + \frac{1}{n-3} = \frac{2}{n}.$$

Multiply throughout by the least common denominator $n(n - 2)(n - 3)$.

Then $\qquad n(n - 3) + n(n - 2) = 2(n - 2)(n - 3).$

$\therefore \qquad n^2 - 3n + n^2 - 2n = 2(n^2 - 5n + 6)$

or $\qquad 2n^2 - 5n = 2n^2 - 10n + 12.$

The terms involving $2n^2$ disappear.

$$\therefore \qquad\qquad\qquad 5n = 12$$

and $\qquad\qquad\qquad\qquad n = \dfrac{12}{5}$.

Exercise 46

Solve the following equations:

1 $\dfrac{18}{2x} - 1 = 2$

2 $\dfrac{4x}{2x - 1} = 5$

3 $\dfrac{x - 1}{x - 2} = 3$

4 $\dfrac{1 - x}{x + 1} = 4$

5 $0.8 = \dfrac{1.5n}{4 + 1.4n}$

6 $\dfrac{5}{2x + 5} = \dfrac{4}{x + 5}$

7 $\dfrac{3p + 23}{3p + 12} = \dfrac{4}{3}$

8 $\dfrac{\frac{1}{x} - 3}{\frac{1}{x} - 2} = \dfrac{1}{4}$

9 $\dfrac{3}{x - 2} - \dfrac{5}{x + 1} = \dfrac{5}{(x - 2)(x + 1)}$

10 $\dfrac{3}{x} - \dfrac{2}{x - 2} = 0$

11 In the formula $\dfrac{1}{f} = \dfrac{1}{v} - \dfrac{1}{u}$ find v when $f = 8$, $u = 2$.

12 For what value of n is $\dfrac{5n - 20}{4n + 6}$ equal to $\frac{1}{3}$?

13 If $C = \dfrac{V}{R}$ find R when $C = 7.5$, $V = 60$.

14 If $r = \dfrac{R(E - V)}{V}$ find V in terms of r, R, E.

16

Graphs of Quadratic Functions

118 Constants and variables

In the formulae which have been considered in earlier chapters it will be seen that the letters employed represent two different kinds of numbers.

(1) Some represent constant numbers – i.e., numbers which remain unchanged in the varying cases to which the formula applies.
(2) Others represent variable quantities.

Example 1: Consider the formula for the circumference of a circle, viz.

$$C = 2\pi r,$$

where   C = the length of the circumference,
               r = the length of the radius.

Two of these four symbols represent constants.

(1) The number 2 (which is fixed for any circle),
(2) The number π (which has always the same value in any formula in which it is used). It represents the constant ratio of the circumference of any circle to its diameter, or the area of a circle to the square of its radius.

Two of the letters in the formula are variables, they are different in value for different circles, the variation of C depending on changes in r.

Example 2: In Section 78 it was shown that the formula, or equation connecting profit and customers, was

$$y = 0.05x - 11.5.$$

It was pointed out that while the profit represented by y depended on the varying number of customers x, the numbers 0.05 and 11.5 remained constant, representing respectively the average amount paid by each customer, and the fixed charges.

Example 3: This last example was a special case of the equation to a straight line, which in Section 84 was shown to be represented in general by the equation:

$$y = mx + b.$$

For a particular straight line, m is constant, representing the gradient of the line, while b is the fixed distance intercepted on the y axis. But x and y vary for different points on the line. They represent the co-ordinates of any point.

119 Dependent and independent variables

The variables are seen to be of two kinds.

Considering the case of the circumference of the circle, the length of this depends on the length of the radius.

In the second example the profit depends on the number of customers.

A variable which thus depends on another variable for its value is called a dependent variable.

The other variable upon which the first one depends is called the independent variable.

As another example, if a train is moving with uniform – i.e., constant – speed, the distance travelled depends on the time.

Thus the speed is a constant,

time is an independent variable,
distance is a dependent variable.

Again the cost of a quantity of tea depends on the weight bought, the price per kilo being constant.

Thus weight is the independent variable.
cost is the dependent variable.

Graphs. In plotting graphs which show how one quantity varies as another varies, the independent variable is always measured on the x axis and the dependent on the y axis.

Not in demand curve

120 Functions

When two quantities are connected as shown, the dependent variable is said to be a function fo the independent variable. Thus:

The circumference of a circle is a function of its radius.

The area of a square is a function of its side.

The distance travelled by a car moving uniformly is a function of the time.

If a spring is stretched by a force, the extension of the spring is a function of the force.

Generally if a quantity denoted by y depends for its value on another quantity x then y is a function of x.

Thus in each of the following examples

$$y = 2x$$
$$y = x^2$$
$$y = \frac{5}{x} - 3$$
$$y = 3(x - 1)(x - 3)$$

the value of y depends on the value of x. If any particular value be given to x, a corresponding value of y can be calculated. In all such cases y is a function of x.

The idea of a function is probably the most important in modern mathematics. From the above examples the following definition can be deduced:

Function. If a quantity y is related to a quantity x so that for every value which might be assigned to x there is a corresponding value of y, then y is a function of x.

Thus for every length that may be chosen for the radius of a circle, there is always a corresponding length of the circumference.

If a man works at a fixed rate per hour, then for any number of hours that he works, there is always a corresponding amount of pay. His pay is a function of the time he works; the rate per hour is a constant.

Again if $y = 2x^3$, then for every value which we may choose to give to x, there is a corresponding value of y. Consequently, y is a function of x.

121 Graph of a function

If y is a function of x, and since, by the above definition, for every value assigned to x there is always a corresponding value of y, then these pairs of values of x and y can be plotted, and the assemblage of points so plotted will be the graph of the function.

Thus every function has a distinctive graph, which will be a regular curve or a straight line, and by which it can be identified.

If the function is of the first degree, and does not involve any algebraical fraction, then, as we have already seen, the graph will be a straight line. For this reason a function of the first degree, of which the general form is $y = mx + b$, is called a linear function.

If, however, the function involves a higher power of x, such as x^2, x^3, etc., or involves an algebraic fraction such as $y = \frac{1}{x}$, the graph will be a regular curve, the shape of which will differ with the nature of the function.

122 Graph of a function of second degree

The simplest form of a function of the second degree is that which is expressed by $y = x^2$. This is called a quadratic function, from the Latin quadratus (squared). The area of a circle, $A = \pi r^2$, is a special form of this.

To plot the curve of $y = x^2$. We first assign values to x, calculate the corresponding values of x^2, or y, and tabulate them as follows:

x	-3	-2.5	-2	-1.5	-1	0	1	1.5	2	2.25	3
y	9	6.25	4	2.25	1	0	1	2.25	4	6.25	9

As wide a range of values is taken as the size of the paper will allow to be plotted. Since the values of y increase more rapidly than those of x, more room is needed on the y axis, but as the square of a number is always positive, no negative values of y are necessary. The x axis is therefore drawn near the bottom of the paper, as shown in Fig. 31.

Selection of scales on axes.   The scale chosen must depend on the number of values we wish to include within the limits of the paper. The same scales need not be chosen on the two axes, and in this particular curve it will be better to take a smaller scale for values of y, because their values increase more rapidly than those of x.

It must be remembered, however, that the true shape of the

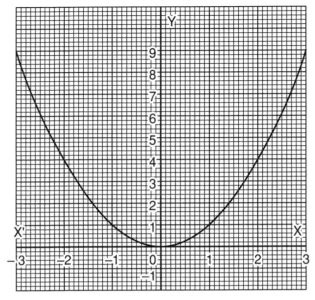

Fig. 31.

curve will be shown only when the same units are taken on both axes.

The symmetry of the values of y, as shown in the table, suggests that the y axis should be in the middle of the paper.

The points when plotted appear to lie on a smooth curve as shown in Fig. 31.

The curve is called a parabola. When inverted it is the curve described by a projectile such as a shot from a gun, or a rocket when fired into the air.

123   Some properties of the curve of $y = x^2$

The following points should be noted about this curve:

(1) The symmetry of the curve.   Positive and negative values of x produce equal values for y. If therefore the curve were folded about the y axis, the two parts of the curve, point by point, would coincide.

   The curve is therefore said to be symmetrical about OY and OY is called the axis of symmetry.

(2) The minimum value of the curve is 0 at the origin. The curve is said to have a turning-point at the origin.

(3) The slope of the curve is not constant, as with a straight line, but increases from point to point as x increases. The gradient is clearly a function of x, since its value depends on the value of x.

(4) The curve may be used to read off the square of any number within the range of plotted values, and also, conversely, to determine square roots. Thus to find $\sqrt{3}$, take the point on the curve corresponding to 3 on the y axis. It is then seen that there are two points on the x axis which correspond to this, the values of x (i.e., $\sqrt{3}$) being $+1.73$ and -1.73 (see Section 49).

124   The curve of $y = -x^2$

All values of y for this curve are equal numerically to the corresponding values of y in $y = x^2$, but are negative. The shape of the curve will be the same, but inverted, as in Fig. 32.

All the values of y being negative, the x axis is drawn toward the top of the paper.

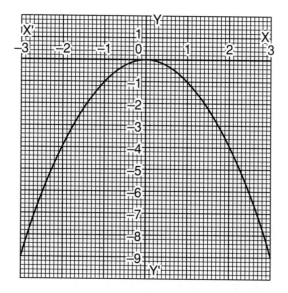

Fig. 32.

The curve has a maximum point – viz. zero – at the origin. From this point the curve shows the path of a bomb dropped from an aeroplane at 0 – ignoring air resistance.

125 The curves of $y = ax^2$

The curves represented by $y = ax^2$, where a is any number, are all parabolas differing from $y = x^2$ only in having different slopes.

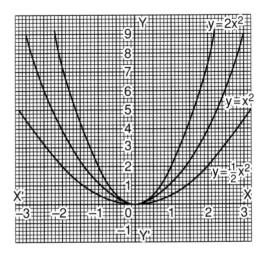

Fig. 33.

Considering $y = 2x^2$ as an example the table of values for plotting the curve are found in the same way as for $y = x^2$.

This curve, as well as that of $y = \frac{1}{2}x^2$, are shown in Fig. 33, and contrasted with $y = x^2$.

If a be negative, we get a corresponding set of curves similar to $y = -x^2$, as in Fig. 32.

126 The curves of $y = x^2 \pm a$, where a is any number

The curve of $y = x^2 + 2$ is evidently related to that of $y = x^2$ in the same way that $y = x + 2$ is connected with $y = x$ (see Section 84). Each ordinate of $x^2 + 2$ is greater by 2 than the corresponding ordinate of x^2.

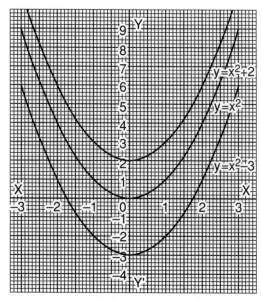

Fig. 34.

The curve of $y = x^2 + 2$ is therefore the same as $y = x^2$ raised two units on the y axis. It appears as shown in Fig. 34. Similarly, the curve $y = x^2 - 3$ is the curve of $y = x^2$, but every point is three units lower than $y = x^2$ for corresponding values of x.

Generally the curves represented by the equation $y = x^2 \pm a$ are a set of curves similar to $y = x^2$ and raised or lowered by an amount equal to $\pm a$, according to the sign of a.

Similarly, the curve of $y = -x^2$ gives rise to a set of curves included in the equation $y = -x^2 \pm a$.

127   Change of axis

On examination of Fig. 34, it will be seen that the curve of $y = x^2 + 2$ differs from the curve of $y = x^2$ only in the position of the x axis. For the curve of $y = x^2 + 2$ the x axis is two units below the lowest point of the curve. Consequently the curve of $y = x^2$ can be changed into that of $y = x^2 + 2$ by drawing a new x axis two units below the original, as shown in Fig. 35, and re-numbering the y axis.

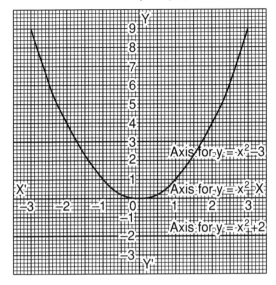

Fig. 35.

Similarly, $y = x^2$ becomes $y = x^2 - 3$ by drawing a new x axis three units above the original, as shown, and again re-numbering the units on the y axis.

128 Curve of $y = (x - 1)^2$

The table below shows the values of (1) $y = (x - 1)^2$ and (2) $y = x^2$, for corresponding values of x.

x	-3	-2	-1	0	1	2	3	4
x^2	9	4	1	0	1	4	9	16
$(x - 1)^2$	16	9	4	1	0	1	4	9

A comparison of the sets of values of two functions shows that both have the same sequence of values, but those of $(x - 1)^2$ are those of x^2 moved one place to the right in the table for consecutive values of x.

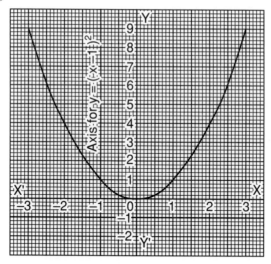

Fig. 36.

Consequently the curve of $(x - 1)^2$ must be the same as that of x^2, but moved one unit to the right.

Therefore, as in the previous paragraph, but moving the y axis instead of the x axis, the curve of x^2 will be changed into the curve of $(x - 1)^2$ by moving the y axis one unit to the left and renumbering the x axis (see Fig. 36).

Similarly if the y axis be moved two units to the left, we get the curve of $(x - 2)^2$. Or if the y axis be moved one unit to the right we should have the curve of $y = (x + 1)^2$, and so for similar functions.

129   Graph of $y = (x - 1)^2 - 4$

By combining the operations illustrated in the preceding two paragraphs – i.e., by moving both axes – the curve of $y = x^2$ can be changed into the curve of such a function as $y = (x - 1)^2 - 4$. This can be done in two steps:

(1) By moving the y axis one unit to the left we get the curve of $y = (x - 1)^2$, as in Section 128.
(2) By now moving the x axis 4 units up the y axis the curve of $y = (x - 1)^2$ becomes $y = (x - 1)^2 - 4$.

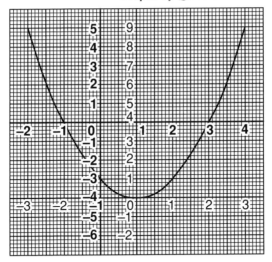

Fig. 37.

With these two new axes, both re-numbered, the curve is as shown in Fig. 37.

In this diagram the new axes have been drawn and the new numbers on the axes marked in larger figures.

The expression may be simplified, since

$$(x - 1)^2 - 4 = x^2 - 2x + 1 - 4$$
$$= x^2 - 2x - 3.$$

The curve in Fig. 37 is therefore the graph of

$$y = x^2 - 2x - 3.$$

We shall see later (see Section 137) that every quadratic function of x can be reduced to such a form as this, or the simpler forms of Sections 127 and 128. We may therefore draw the conclusion that:

The graph of every quadratic function of x is a modified form of $y = x^2$ and is therefore a parabola.

130   The graph $y = x^2 - 2x - 3$

We have seen in the preceding paragraph that this graph can be obtained from the curve of $y = x^2$ by changing axes. In practice, however, the method given below will be found more convenient in most cases and usually more accurate in practical application. A table of values is constructed as follows:

x	− 3	− 2	− 1	0	1	2	3	4
x^2	9	4	1	0	1	4	6	16
− 2x	+ 6	+ 4	+ 2	0	− 2	− 4	− 6	− 8
− 3	− 3	− 3	− 3	− 3	− 3	− 3	− 3	− 3
y	12	5	0	− 3	− 4	− 3	0	5

Note: Be careful not to add the value of x in each column. It is well to draw a thick line under the values as shown, to remind one when adding.

From the values in this table you should plot the curve and compare the figure obtained with that in Fig. 37, arrived at by change of axes. It will be seen that the minimum value of $x^2 − 2x − 3$ is − 4, when x = 1, and the line x = 1, perpendicular to the x axis, is an axis of symmetry.

131 Solution of the equation $x^2 − 2x − 3 = 0$ from the graph

At the points where the curve of $x^2 − 2x − 3$, in Fig. 37, cuts the axis of x, the value of y is zero − i.e.:

$$x^2 − 2x − 3 = 0.$$

The values of x at these two points are:

$$x = 3 \text{ and } x = −1.$$

We have thus found two values of x which satisfy the equation $x^2 − 2x − 3 = 0$. The equation has therefore been solved by means of the graph.

132 Graph of $y = 2x^2 − 3x − 5$

This is a slightly more difficult graph to plot. The table of values is as follows:

x	− 3	− 2	− 1	0	$\frac{1}{2}$	1	2	3	4	5
$2x^2$	18	8	2	0	$\frac{1}{2}$	2	8	18	32	50
− 3x	9	6	3	0	$-1\frac{1}{2}$	− 3	− 6	− 9	− 12	− 15
− 5	− 5	− 5	− 5	− 5	− 5	− 5	− 5	− 5	− 5	− 5
y	22	9	0	− 5	− 6	− 6	− 3	4	15	30

In this example a wider range of values is shown. As the values of y increase rapidly smaller units are taken on the two axes. The graph is as shown in Fig. 38.

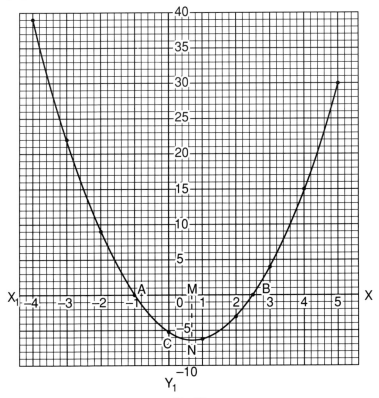

Fig. 38.

The lowest point, N, giving the minimum value of the expression corresponds to $x = \frac{3}{4}$. From the table of values this point is seen to be half-way between $x = \frac{1}{2}$ and $x = 1$, since these values of x give the same values of y – viz., -6. The minimum value represented by MN, is $-6\frac{1}{8}$. The ordinate through M, of which MN is part, is the axis of symmetry of the curve.

The solution of the equation:

$$2x^2 - 3x - 5 = 0,$$

will be given by the values of x, where $y = 0$ – i.e., at A and B, where the curve cuts the axes. These points give:

$$x = -1, \text{ or } 2.5.$$

133   Graph of $y = 12 - x - x^2$

In this example the coefficient of x^2 is negative. Consequently the curve will take the form of $y = -x^2$, as shown in Fig. 32 – i.e., it will be an inverted parabola. The table of values is as follows:

x	-5	-4	-3	-2	-1	0	1	2	3	4
12	12	12	12	12	12	12	12	12	12	12
$-x$	5	4	3	2	1	0	-1	-2	-3	-4
$-x^2$	-25	-16	-9	-4	-1	0	-1	-4	-9	-16
y	-8	0	6	10	12	12	10	6	0	-8

The values of y show the symmetry of the curve. The highest point, giving the maximum value, is seen to be half-way between $x = -1$, and $x = 0$ – i.e., where $x = -\frac{1}{2}$. The curve is as shown in Fig. 39. The maximum value is at M, where the value of y is $11\frac{1}{4}$.

The ordinate from M – i.e., MN – is the axis of symmetry for the curve.

We can obtain from the curve the solution of the equation:

$$12 - x - x^2 = 0 \text{ – i.e. } x^2 + x - 12 = 0.$$

This will be given by the values of x at A and B, where the curve cuts the x axis. The solutions are:

$$x = 3, \ x = -4.$$

Fig. 39.

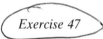

Exercise 47

1 Draw the curve of $y = x^2$ between the values $x = +3$ and $x = -3$, taking the units as large as possible. From the curve write down the values of:

    (1) 2.3^2 (2) $\sqrt{7}$ (3) $\sqrt{3.5}$

2 Draw the curve of $y = \frac{1}{4}x^2$ between $x = +4$ and $x = -4$. Find from the curve the values of x such that:

    (1) $\frac{1}{4}x^2 = 0.8$ (2) $\frac{1}{4}x^2 = 2$ (3) $x^2 = 12$

3 Draw the curve of $-\frac{1}{2}x^2$ between $x = +4$ and $x = -4$. From the curve find the values of:

    (1) $-\frac{1}{2}x^2 = -3$ (2) $-x^2 = -5$ (3) $x^2 = 2$

4 Draw the curve of $y = x^2$ and by change of axes obtain the curve of $y = x^2 + 3$. From the curve find the values of x such that:

    (1) $x^2 + 3 = 5$ (2) $x^2 + 3 = 9$

5 Draw the curve of $y = x^2$ and by change of axes obtain the curves of:

    (*a*) $y = (x - 2)^2$ (*b*) $y = (x + 3)^2$

From the curves find the values of x such that:

(1) $(x - 2)^2 = 3$ (2) $(x + 3)^2 = 1$
(3) $(x - 2)^2 - 5 = 0$

6   Draw the curve of $y = (x + 2)^2 - 2$ by means of the curve of $y = x^2$ and changes of axes. Use the curve to find the values of x when:

(1) $(x + 2)^2 - 2 = 0$ (2) $(x + 2)^2 - 1 = 0$
(3) $(x + 2)^2 = 8$

7   Draw the curve of $y = x^2 - 6x + 5$. Find the least value of this function and the corresponding value of x. Use the curve to find the values of x when:

(1) $x^2 - 6x + 5 = 0$ (2) $x^2 - 6x + 5 = 6$

8   Draw the curve of $y = x^2 - 4x + 2$. Find the minimum value of the function and the corresponding value of x. Use the curve to solve the following equations:

(1) $x^2 - 4x - 2 = 0$ (2) $x^2 - 4x - 2 = 3$

9   Draw the curve of $y = 2x^2 - 5x + 2$. Find the minimum value of the function and the corresponding value of x. From the curve solve the equations:

(1) $2x^2 - 5x + 2 = 0$ (2) $2x^2 - 5x - 1 = 0$

10   Draw the curve of $y = 2 - x - x^2$. Find the maximum value of the function and the corresponding value of x.

134   Using graphs to solve quadratic inequalities

Once we have drawn the graph of a quadratic function we can use it to write down the solutions for an associated quadratic inequality.

Example: By drawing the graph of $y = x^2 - 4x + 3$ find the values of x for which $x^2 - 4x + 3 > 0$.

The values of x for which $x^2 - 4x + 3 > 0$ correspond to the places on the graph where $y > 0$. These are where y is positive or above the x-axis. The values of x where y is above the axis are where $x < 1$ or where $x > 3$. These values can be shown using square brackets i.e. $]-\infty, 1[$ or $]3, \infty[$.

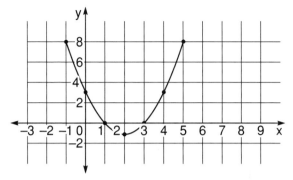

The values of x for which $x^2 - 4x + 3 < 0$ correspond to the places on the graph where $y < 0$. These are where y is negative or below the x-axis. The values of x, where y is below the axis, are between $x = 1$ and $x = 3$ i.e. $]1, 3[$.

Exercise 48

1 Draw the graph of $y = x^2 - 3x + 2$.
Write down the values of x where the graph lies above the x-axis. Hence write down the values of x for which:

(a) $x^2 - 3x + 2 > 0$ (b) $x^2 - 3x + 2 < 0$

2 Draw the graph of $y = x^2 + 2x - 3$.
Write down the values of x where the graph lies above the x-axis. Hence write down the values of x for which:

(a) $x^2 + 2x - 3 \geqslant 0$ (b) $x^2 + 2x - 3 < 0$

3 By drawing the graph of $y = x^2 - 3x - 4$ find the values of x for which $x^2 - 3x - 4 > 0$.

4 By drawing the graph of $y = x^2 + 4x - 5$ find the values of x for which $x^2 + 4x - 5 \leqslant 0$.

135 Using quadratic inequalities to describe regions

We saw in Chapter 12 that $y > x + 2$ and $y < x + 2$ represented the regions on either side of the line $y = x + 2$. A similar situation arises whereby $y > x^2 - 5x + 4$ and $y < x^2 - 5x + 4$ represent the regions above and below the curve $y = x^2 - 5x + 4$.

Example: Draw the graph of $y = x^2 - 5x + 4$ and show the regions for which (a) $x^2 - 5x + 4 > 0$ and (b) $x^2 - 5x + 4 < 0$.

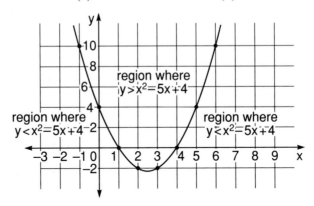

Note: the point (2, 3) lies in the region above the curve $y = x^2 - 5x + 4$ and when substituting the coordinates of this point we should find that the y coordinate, 3, *is greater than* $2^2 - 5 \times 2 + 4 = -2$ ($3 > -2$).

The point $(-1, 5)$ lies in the region below the curve and when substituting we should find that the y value, 5, *is less than* $(-1)^2 - 5 \times (-1) + 4 = 10$.

Exercise 49

1   Draw the graph of $y = x^2 - 4x + 3$ and show the regions for which:

(a) $x^2 - 4x + 3 > 0$ and (b) $x^2 - 4x + 3 < 0$

Check that the point (1, 5) lies in the region above the graph.

2   Draw the graph of $y = x^2 - 4x - 5$ and show the regions for which:

(a) $x^2 - 4x - 5 > 0$ and (b) $x^2 - 4x - 5 < 0$

Check that the point (1, 5) lies in the region above the graph.

3   Draw the graphs of $y = x^2$ and $y = 2x + 3$.
Shade the region where $y > x^2$ and $y < 2x + 3$.
Check that the point (1, 3) lies in this region.

4   Draw the graphs of $y = x^2 - 3x$ and $y = 2x - 4$.
Shade the region where $y < x^2 - 3x$ and $y > 2x - 4$.
Find a point in the region where $y > x^2 - 3x$, $y < 2x - 4$ and $y > 0$.

17

Quadratic Equations

136 Algebraical solution

Plotting the graph of a quadratic function led logically to the solution of a quadratic equation – i.e., an equation of the second degree. The solution by this method is useful and illuminating, but as a method of solving a quadratic equation it is cumbersome, and the accuracy obtainable is limited. We must therefore find an algebraical solution which is certain, universally applicable and capable of any required degree of accuracy.

A quadratic equation was solved for the first time when, in Section 123, from the curve of $y = x^2$ it was found that if $x^2 = 3$, the corresponding values of x were $+1.73$ and -1.73. The reasoning may be stated thus:

$$x^2 = 3.$$
$$\therefore \qquad x = \pm \sqrt{3}$$
$$\text{and} \qquad x = +1.73 \text{ or } -1.73.$$

This is the simplest form of a quadratic equation. It involves the operation of finding a square root; hence the term 'root' as applied to the solution of an equation.

In Section 129 an important step forward was made. The points where the curve of $(x - 1)^2 - 4$ cuts the x axis, and the function was therefore equal to zero, were found, and the corresponding values of x noted – viz., 3 and -1.

This means that for these values of x, the expression

$$(x - 1)^2 - 4 = 0.$$

They are therefore the roots of this equation.

Let the equation be written in the form:

$$(x - 1)^2 = 4 \qquad\qquad (A)$$

Then algebraically it is of the same form as the equation above, $x^2 = 3$. We can proceed with the algebraical solution on the same lines.

Taking the square roots of each side:

$$x - 1 = \pm 2.$$
$$\therefore \qquad x - 1 = + 2 \text{ or } x - 1 = -2,$$
whence $\qquad\qquad x = 3 \text{ or } x = -1.$

The form marked (A) is the form to which, ultimately, all quadratic equations are reduced; our object always is to reach this form.

137 The method of solution of any quadratic

It was shown that the expression $(x - 1)^2 - 4$ simplified to $x^2 - 2x - 3$.

∴ the equation which was solved could have been written in the form:

$$x^2 - 2x - 3 = 0 \qquad\qquad (1)$$

If we start with this equation and wish to solve it, we need to get back to the form:

$$(x - 1)^2 - 4 = 0 \qquad\qquad (2)$$

This is the converse operation of changing from (2) to (1), i.e. of obtaining the complete square (2) when we are given (1).

~~Two~~ 3 preliminary steps are necessary.

(1) Remove the constant to the right side, as it does not help in finding the square.
(2) Divide throughout by the coefficient of x^2, if this is not unity.

Finally we arrive at the form:

$$x^2 - 2x = 3.$$

It is now necessary to add to the left side such a number as will produce a complete square. Remembering the work in Section 106, we get the following rule:

(3) Add to each side the square of half the coefficient of x.

Then the above becomes

$$x^2 - 2x + (1)^2 = 3 + 1.$$
$$\therefore \qquad (x - 1)^2 = 4,$$

and we proceed as before.

We will now apply the above method to solve the equation:

$$x^2 - 6x + 5 = 0.$$
Then $\qquad x^2 - 6x = -5.$

Add to each side $(\frac{6}{2})^2$ – i.e., 3^2.

Then $\qquad x^2 - 6x + (3)^2 = -5 + 9.$
$$\therefore \qquad (x - 3)^2 = 4.$$

Thus we reach the desired form. Proceeding as before:

$$x - 3 = \pm 2.$$
$$\therefore \qquad x = 3 \pm 2.$$
Then $\qquad x = 3 + 2 = 5$
or $\qquad x = 3 - 2 = 1.$

$\therefore$ the solution is $x = 5$ or $x = 1$.

138   Solution of $2x^2 + 5x - 3 = 0$

Applying the preliminary steps (1) and (2) of the previous paragraph, we get in succession:

$$2x^2 + 5x = 3$$
and $\qquad x^2 + \frac{5}{2}x = \frac{3}{2}.$

Half the coefficient of x is $\frac{5}{4}$.
Adding $(\frac{5}{4})^2$ we get:

$$x^2 + \frac{5}{2}x + (\frac{5}{4})^2 = \frac{3}{2} + (\frac{5}{4})^2.$$
$$\therefore \qquad (x + \frac{5}{4})^2 = \frac{3}{2} + \frac{25}{16} = \frac{49}{16}.$$

Taking square roots of both sides:

$$x + \frac{5}{4} = \pm \frac{7}{4}.$$
$$\therefore \qquad x = -\frac{5}{4} \pm \frac{7}{4}$$
and $\qquad x = -\frac{5}{4} + \frac{7}{4} = \frac{2}{4}$
or $\qquad x = -\frac{5}{4} - \frac{7}{4} = -\frac{12}{4}.$
$$\therefore \qquad x = \frac{1}{2} \text{ or } x = -3.$$

139 Worked examples

Example 1: Solve the equation $x^2 - x - 1 = 0$.

Transposing

$$x^2 - x = 1.$$

Adding $(\frac{1}{2})^2$

$$x^2 - x + (\tfrac{1}{2})^2 = 1 + \tfrac{1}{4}$$
or
$$(x - \tfrac{1}{2})^2 = \tfrac{5}{4}.$$
$$\therefore \qquad x - \tfrac{1}{2} = \pm\frac{\sqrt{5}}{2}$$
and
$$x = \tfrac{1}{2} \pm \frac{\sqrt{5}}{2}\ .$$
$$\therefore \qquad x = \tfrac{1}{2} + \frac{\sqrt{5}}{2} = \frac{1 + 2.236}{2} = 1.618$$
or
$$x = \tfrac{1}{2} - \frac{\sqrt{5}}{2} = \frac{1 - 2.236}{2} = -0.618$$
$$\therefore \qquad x = 1.618 \text{ or } x = -0.618 \ \text{(both approx.)}.$$

Example 2: Solve the equation $3x^2 - 5x + 1 = 0$.

Applying preliminary steps, Section 137:

$$3x^2 - 5x = -1$$
$$x^2 - \tfrac{5}{3}x = -\tfrac{1}{3}.$$

Adding $(\frac{5}{6})^2$:

$$x^2 - \tfrac{5}{8}x + (\tfrac{5}{6})^2 = -\tfrac{1}{3} + \tfrac{25}{36}.$$
$$\therefore \qquad (x - \tfrac{5}{6})^2 = \tfrac{13}{36}.$$

Taking square roots

$$x - \tfrac{5}{6} = \frac{\pm 3.606 \ \text{(approx.)}}{6}$$
$$\therefore \qquad x = \frac{5 \pm 3.606}{6}\ .$$
$$\therefore \qquad x = \frac{5 + 3.606}{6} = \frac{8.606}{6} = 1.434$$
or
$$x = \frac{5 - 3.606}{6} = \frac{1.394}{6} = 0.232.$$
$$\therefore \text{ solution is} \qquad x = 1.434 \text{ or } x = 0.232 \ \text{(approx.)}.$$

Example 3: Solve the equation:

$$\frac{1}{x - 1} - \frac{1}{x + 2} = \frac{1}{16}\ .$$

First we clear fractions by multiplying throughout by the least common denominator – viz., $16(x - 1)(x + 2)$.

Then $\quad 16(x + 2) - 16(x - 1) = (x - 1)(x + 2)$.
$\therefore \qquad 16x + 32 - 16x + 16 = x^2 + x - 2$.

Adding and transposing:
$$-x^2 - x = -48 - 2$$
or $\qquad x^2 + x = 50$.

Adding $(\frac{1}{2})^2$:

$$x^2 + x + (\tfrac{1}{2})^2 = 50 + \tfrac{1}{4} = \tfrac{201}{4}.$$
$$\therefore \qquad (x + \tfrac{1}{2})^2 = \pm \frac{\sqrt{201}}{2} = \frac{\pm 14.177 \text{ (approx.)}}{2}.$$

$$\therefore \qquad x = -\tfrac{1}{2} \pm \frac{14.177}{2}$$

and $\qquad x = -\tfrac{1}{2} + \dfrac{14.177}{2} = \dfrac{13.177}{2} = 6.588$

or $\qquad x = -\tfrac{1}{2} - \dfrac{14.177}{2} = -\dfrac{15.177}{2} = -7.588$.

$\therefore$ the solution is
$$x = 6.588 \text{ or } x = -7.588.$$

Exercise 50

Solve the following equations:

1 $3x^2 = 12$ $\qquad\qquad$ **2** $4x^2 - 1 = 0$

3 $\dfrac{x^2}{16} = 9$ $\qquad\qquad$ **4** $(x + 1)^2 - 4 = 0$

5 $(x - 3)^2 - 25 = 0$ $\qquad$ **6** $(x + 5)^2 = 36$

7 $(x + \tfrac{1}{4})^2 = 1$ $\qquad\qquad$ **8** $\dfrac{(x - 9)^2}{15} = 4$

9 $x^2 - 10x + 16 = 0$ $\qquad$ **10** $x^2 + x - 12 = 0$
11 $x^2 - 2x - 15 = 0$ $\qquad$ **12** $x^2 + 3x - 28 = 0$
13 $x(x - 4) = 32$ $\qquad\qquad$ **14** $2x^2 - 7x + 6 = 0$
15 $2x^2 - 3x - 5 = 0$ $\qquad$ **16** $3x^2 = 7x + 9$

17 $3x^2 + 1 = 5x$ $\qquad\qquad$ **18** $\dfrac{6}{x} + \dfrac{3}{x + 2} = 2$

19 $3x - \dfrac{5}{x} = 14$ $\qquad\qquad$ **20** $x - 2 = \dfrac{2}{x}$

21 $\dfrac{x - 9}{3} = \dfrac{x + 5}{x}$ $\qquad$ **22** $\dfrac{1}{x - 1} - \dfrac{1}{x + 2} = \dfrac{1}{16}$

140 Solution of quadratic equations by factorisation

There is another method of solving quadratics; it is as follows:
We first note that if n be any finite number

$$n \times 0 = 0,$$

i.e., the product of any finite number and zero is always zero.

Conversely, if the product of two factors is zero, then either of the factors may be zero.

For example, if $a \times b = 0$ then this is true if either $a = 0$, or $b = 0$.

Similarly if

$$(x - 1)(x - 3) = 0 \qquad \text{(A)}$$

it follows from the above that this equation is satisfied if either $x - 1 = 0$ or $x - 3 = 0$.

But if $x - 1 = 0$, then $x = 1$, and if $x - 3 = 0$, then $x = 3$.

∴ equation (A) muist be satisfied if either $x = 1$ or $x = 3$.

∴ 1, and 3, are the roots of the equation:

$$(x - 1)(x - 3) = 0,$$
i.e.,
$$x^2 - 4x + 3 = 0.$$

If it is required to solve the equation $x^2 - 4x + 3 = 0$, this can be done by reversing the above steps; consequently we find the factors of $x^2 - 4x + 3$ and so we get:

$$(x - 1)(x - 3) = 0.$$

The method is an easy one, if it is possible to obtain factors. In equations which arise out of practical work this can seldom be done.

It is a valuable method in other ways, and can be employed in equations of higher degree. If, for example, we know that

$$(x - 1)(x - 3)(x - 4) = 0.$$

Then by the above reasoning this equation is satisfied by:

$$x - 1 = 0, \text{ whence } x = 1$$
$$x - 3 = 0, \text{ whence } x = 3$$
$$x - 4 = 0, \text{ whence } x = 4.$$

The product of the three factors is an expression of the third degree, since the term of highest degree will be x^3. The equation is therefore of the third degree, or a cubic equation.

141   Worked examples

Example 1: Solve the equation $x^2 - 2x = 15$.

Transposing
$$x^2 - 2x - 15 = 0.$$

Factorising
$$(x - 5)(x + 3) = 0.$$
$$\therefore \qquad\qquad x - 5 = 0 \text{ and } x = 5$$
or $\qquad\qquad x + 3 = 0 \text{ and } x = -3.$

$\therefore$ the solution is $x = 5$ or $x - -3$.

Note: You should remember that the reasons for this method require that the right-hand side should be zero. Thus it cannot be used in such a case as

$$(x - 6)(x - 2) = 4.$$

Example 2: Solve the equation $9x(x + 1) = 4$.

Simplifying and transposing
$$9x^2 + 9x - 4 = 0.$$

Factorising
$$(3x + 4)(3x - 1) = 0.$$
$$\therefore \qquad\qquad 3x + 4 = 0 \text{ and } x = -\tfrac{4}{3}$$
or $\qquad\qquad 3x - 1 = 0 \text{ and } x = \tfrac{1}{3}.$

$\therefore$ the solution is $x = -\tfrac{4}{3}$ or $x = \tfrac{1}{3}$.

Exercise 51

Solve the following equations by the method of factors:

1	$x(x - 3) = 0$	**2**	$x(x + 5) = 0$
3	$(x - 2)^2 = 0$	**4**	$(x - 1)(x - 2) = 0$
5	$(x + 4)(x - 1) = 0$	**6**	$2(x - 3)(x + 7) = 0$
7	$4(3x - 7)(2x + 11) = 0$	**8**	$x^2 - 9x + 20 = 0$
9	$x^2 + x = 6$	**10**	$x^2 + 2x = 35$
11	$x(x + 13) + 30 = 0$	**12**	$x(x - 4) = x + 66$
13	$(x - 8)(x + 4) = 13$	**14**	$2x^2 - 11x + 12 = 0$
15	$2x^2 - 3x - 5 = 0$	**16**	$3x^2 - 4x + 1 = 0$
17	$8x^2 - 14x - 15 = 0$	**18**	$24x^2 + 10x - 4 = 0$

19 $(x - 2)(x - 4)(x - 5) = 0$
20 $(2x - 1)(x + 3)(x + 2) = 0$
21 $(x - 1)(x^2 - 2x - 8) = 0$
22 $(2x - 5)(x^2 - 5x - 50) = 0$
23 $(x - a)(x - b) = 0$ **24** $(2x - c)(x + d) = 0$

142   General formula for the solution of a quadratic equation

We have seen that by simplification and transposing every quadratic can be written in a form such as:

$$2x^2 + 7x - 4 = 0,$$

in which there are three constants – viz.:

(1) the coefficient of x^2,
(2) the coefficient of x,
(3) the term independent of x.

   Consequently if we want to write down a general form for any quadratic, letters such as a, b, c can be chosen to represent these constants, so that we could write the general form as:

$$ax^2 + bx + c = 0.$$

The equation above is a special case of this in which $a = 2$, $b = 7$, $c = -4$.

   If we solved the general quadratic, $ax^2 + bx + c = 0$, the roots would be in terms of a, b, c. We should then have a formula such that, by substituting the values of a, b and c in any special case, we should be able to write down the roots, and no actual solving would be necessary.

143   Solution of the quadratic equation $ax^2 + bx + c = 0$

We must use the method of 'completing the square' as in Section 137:

$$ax^2 + bx + c = 0.$$

Using preliminary step (1) $ax^2 + bx = -c.$

Using preliminary step (2) $x^2 + \dfrac{b}{a} x = -\dfrac{c}{a}.$

Adding 'the square of half the coefficient of x'

$$x^2 + \frac{b}{a}x + \left(\frac{b}{2a}\right)^2 = \frac{b^2}{4a^2} - \frac{c}{a}.$$

$\therefore \qquad \left(x + \frac{b}{2a}\right)^2 = \frac{b^2 - 4ac}{4a^2}$ (adding the fractions).

Taking square roots

$$x + \frac{b}{2a} = \pm \frac{\sqrt{b^2 - 4ac}}{2a}.$$

$\therefore \qquad x = -\frac{b}{2a} \pm \frac{\sqrt{b^2 - 4ac}}{2a}$

or $\qquad x = \frac{-b \pm \sqrt{b^2 - 4ac}}{2a}$

or in full $\qquad x = \frac{-b + \sqrt{b^2 - 4ac}}{2a}$

or $\qquad x = \frac{-b - \sqrt{b^2 - 4ac}}{2a}.$

This formula should be carefully learnt: by using it, there is no real necessity for working out any quadratic equation. Before applying it, however, the equation to be solved must be written down with all the terms on the left side.

From this formula, since there are always two square roots of a number, $\sqrt{b^2 - 4ac}$ must always have two values. Therefore x must have two values and every quadratic equation must have two roots. This fact was obvious in the solution by means of a graph.

144 Worked examples

Example 1: Solve the equation $5x^2 + 9x = 2$.

Writing the equation in general form we get:

$$5x^2 + 9x - 2 = 0.$$

Using the formula

$$x = \frac{-b \pm \sqrt{b^2 - 4ac}}{2a}$$

we have for this example

$$a = 5, b = 9, c = -2.$$

Substituting

$$x = \frac{-9 \pm \sqrt{9^2 - 4 \times 5 \times -2)}}{2 \times 5}$$

$$= \frac{-9 \pm \sqrt{81 + 40}}{10}$$

$$= \frac{-9 \pm \sqrt{121}}{10}$$

$$= \frac{-9 \pm 11}{10}$$

$$\therefore \qquad x = \frac{-9 + 11}{10} = \frac{2}{10}$$

$$\text{or} \qquad x = \frac{-9 - 11}{10} = \frac{-20}{10}$$

$$\therefore \qquad x = 0.2 \text{ or } -2.$$

Example 2: Solve the equation $\dfrac{1}{x - 1} + \dfrac{2}{3} = \dfrac{2}{x - 3}$.

Clearing fractions by multiplying throughout by $3(x - 1)(x - 3)$

$$3(x - 3) + 2(x - 1)(x - 3) = 6(x - 1)$$
$$3x - 9 + 2(x^2 - 4x + 3) = 6x - 6$$
$$3x - 9 + 2x^2 - 8x + 6 = 6x - 6.$$

Collecting like terms

$$2x^2 - 11x + 3 = 0.$$

Using the formula

$$x = \frac{-b \pm \sqrt{b^2 - 4ac}}{2a}$$

and putting $a = 2, b = -11, c = 3,$

we get $\qquad x = \dfrac{11 \pm \sqrt{121 - 24}}{4}$

$$= \frac{11 \pm \sqrt{97}}{4} = \frac{11 + 9.849 \text{ (app.)}}{4}.$$

$$\therefore \qquad x = \frac{11 + 9.849}{4} = \frac{20.849}{4} = 5.212$$

$$\text{or} \qquad x = \frac{11 - 9.849}{4} = \frac{1.151}{4} = 0.288$$

$\therefore$ the solution is $x = 5.212$ or $x = 0.288.$

Exercise 52

Solve the following equations by using the formula of Section 143:

1 $x^2 + 3x - 1 = 0$ **2** $x^2 - 5x + 2 = 0$

3 $x^2 - 0.4x = 1.6$ **4** $3x^2 - 5x + 1 = 0$

5 $2x^2 - 5x = 2$ **6** $5x^2 = 7x + 3$

7 $2x^2 + 12x - 7 = 0$ **8** $3x^2 = 8x + 8$

9 $4x^2 + 10x = 5$ **10** $4x(x + 2) = 9$

11 $4x^2 = (x + 4)(2 - x)$ **12** $0.9(x + 1) = 0.8 - x^2$

13 $3x^2 = 7x + 2$ **14** $\dfrac{2}{3x + 1} + \dfrac{1}{2x + 1} = \dfrac{2}{7}$

15 $0.5 + \dfrac{2}{x + 3} - \dfrac{11}{x + 7} = 0$ **16** $\dfrac{3}{x + 2} - \dfrac{1}{x - 3} = \dfrac{4}{x}$

17 $(x + 3)(x - 5) = 2x - 2$

---

145   Problems leading to quadratics

The following examples illustrate the method of solving such problems.

Example 1: The distance (h) which a body reaches in time t when it is projected vertically upwards with velocity u is given by the formula

$$h = ut - \tfrac{1}{2}gt^2.$$

If $u = 160$ and $g = 10$, find the time a body takes to rise 240.

Substituting the given values in the formula

$$240 = 160t - 5t^2$$
$$5t^2 - 160t + 240 = 0$$
$$t^2 - 32t + 48 = 0$$

$$\therefore \quad t = \frac{32 \pm \sqrt{32^2 - 4 \times 48}}{2}$$
$$= 16 \pm 4\sqrt{13}$$
$$= 16 \pm 3.6 \text{ approx.}$$
$$t = \cancel{19.6} \text{ or } \cancel{12.4}. \quad 1.58 \quad 30.32$$

The two roots require consideration. Since every quadratic has two roots, when these furnish the answer to a problem the applicability of the roots to the problem must be examined. Sometimes it will be clear that both are not applicable, especially if one is negative. Considering the above problem, when a body is projected vertically upwards its velocity decreases until it reaches its highest point, when it is zero. It then falls vertically and retraces its path. Therefore it will be at a given height twice, once when ascending and again when descending. The value t = 1·84 gives the time to reach 204 ft. when going up, and it is at the same height when descending 8·16 secs. after starting.

Example 2: A motorist travels a distance of 84 km. He finds that, if on the return journey he increases his average speed by 4 km/h, he will take half an hour less. What was his average speed for the first part of the journey and how long did he take for the double journey?

Let x miles per hour be the average speed for the first journey.

Then time for the first journey is $\dfrac{84}{x}$ hours.

Speed for the return journey is $(x + 4)$ miles per hour.

∴ time for the second journey is $\dfrac{84}{x + 4}$.

But this is $\frac{1}{2}$ hour less than the first time.

∴ $\qquad\qquad \dfrac{84}{x} - \dfrac{84}{x + 4} = \frac{1}{2}$.

Clearing fractions by multiplying by the L.C.D. – viz., $2x(x + 4)$.

$$168(x + 4) - 168x = x(x + 4)$$
$$168x + 672 - 168x = x^2 + 4x.$$

Collecting and transposing

$$x^2 + 4x - 672 = 0.$$

Using the formula

$$x = \frac{-b \pm \sqrt{b^2 - 4ac}}{2a}$$

or substituting $\quad x = \dfrac{-4 \pm \sqrt{16 - (4 \times -672)}}{2}$

$$= \frac{-4 \pm \sqrt{2704}}{2}$$

$$= \frac{-4 \pm 52}{2}.$$

∴ $\qquad\qquad x = \dfrac{-4 + 52}{2} = 24$

or $\qquad\qquad x = \dfrac{-4 - 52}{2} = -28.$

The negative root, although it satisfies the equation above, has no meaning for this problem.

$\therefore$ average speed for first journey is 24 km/h.

Time for first journey $\quad = \frac{84}{24} = 3\frac{1}{2}$ h.
Time for second journey $= \frac{84}{28} = 3$ h.
$\therefore$ total time $= 6\frac{1}{2}$ hours.

Exercise 53

1 The sum of a number and its reciprocal is 2.9. Find the number.

2 The area of a rectangle is 135 mm^2 and its perimeter is 48 mm. What are the lengths of its sides?

3 Solve for $\frac{1}{R}$ the equation $\frac{1}{R^2} - \frac{5}{R} = 10$.

4 The relation between the joint resistance R and two resistances r_1 and r_2 is given by the formula $\frac{1}{R} = \frac{1}{r_1} + \frac{1}{r_2}$. If R is 12 ohms and r_2 is 6 ohms greater than r_1, find r_1 and r_2.

5 A formula for finding the strength of a concrete beam is $bn^2 + 2am(n - c) = 0$. Solve this for n when b = 4, a = 2, c = 8, m = 12.5.

6 The formula giving the sag (D) in a cable of length L and span S is expressed by $L = \frac{8D^2}{3S} + S$. Find S when L = 80, D = 2.5.

7 The product of a number n and 2n − 5 is equal to 250. What is the value of n?

8 There is an algebraical formula $s = \frac{n}{2}\{2a + (n - 1)d\}$. If s = 140, a = 7 and d = 3, find n.

9 By a well-known geometrical theorem the square on the diagonal of a rectangle is equal to the sum of the squares on the two sides. The diagonal of a particular rectangle is 25 mm long, and one side of the rectangle is 5 mm longer than the other. Find the sides of the rectangle.

10 The cost of a square carpet is £24.80 and the cost of another square carpet whose side is 3 m longer than that of the first is £38.75. If the cost per m^2 is the same for both carpets, find the area of each.

11 There is a number such that when it is increased by 3 and the sum is squared, the result is equal to 12 times the number increased by 16. What is the number?

12 Two adjacent sides of a rectangle are represented by $(x + 4)$ and $(x + 6)$. The area of the rectangle is equal to twice the area of the square whose side is x. What is the value of x?

13 Find the area of a rectangular plot of ground whose perimeter is 42 m and whose diagonal is 15 m.

14 The cost of boring a well is given by the formula, $C = 2x + \dfrac{1}{50} x^2$, where C is the cost in pounds and x is the depth in metres. If a well cost £280 to bore, how deep was it?

15 One number exceeds another by 4. The sum of their squares is 208. What are the numbers?

16 The formula for the sum of the first n whole numbers is $\frac{1}{2}n(n + 1)$. If the sum is 78, how many numbers are there?

146　Simultaneous equations of the second degree

These are equations which involve two unknowns and which include terms of the second degree.

They are also called simultaneous quadratics.

The degree of a term is shown either by its index, or, if it contains two or more letters, by the sum of their indices.

x^2, xy, y^2 are terms of the second degree involving two unknowns, and any of these, together with terms of the first degree, may occur in the equations to be solved. Numerical coefficients do not affect the 'degree' of a term.

It is seldom possible to solve simultaneous quadratics unless they conform to certain types. In this book we shall consider two of these types.

147　First type.　When one of the equations is of the first degree

Example

$$x + y = 1 \qquad (1)$$
$$3x^2 - xy + y^2 = 37 \qquad (2)$$

It is always possible to solve simultaneous quadratics if we can find one letter in terms of the other, the relation being a linear one.

In the equations above, from (1) we get:

$$y = 1 - x.$$

This can now be substituted in equation (2).

Thus we get:

$$3x^2 - x(1 - x) + (1 - x)^2 = 37.$$

In this way a quadratic, with one unknown, is reached, and this can always be solved by the methods previously given.

Simplifying

$$3x^2 - x + x^2 + 1 - 2x + x^2 = 37.$$
$$\therefore \qquad 5x^2 - 3x - 36 = 0.$$

This could be solved by factorisation, or using the formula:

$$x = \frac{-b \pm \sqrt{b^2 - 4ac}}{2a}.$$

On substituting $x = \dfrac{+3 \pm \sqrt{9 + 720}}{10}$

$$= \frac{3 \pm 27}{19}.$$

$$\therefore \qquad x = \frac{30}{10} \text{ or } \frac{24}{10}$$

or $$x = 3 \text{ or } -2.4.$$

To find y substitute in

$$y = 1 - x.$$

(1) When $x = 3$,

$$y = 1 - 3 = -2.$$

(2) When $x = -2.4$,

$$y = 1 - (-2.4) = 3.4.$$

$\therefore$ the solutions are:

(1) $x = 3, y = -2.$
(2) $x = -2.4, y = 3.4.$

The solutions should be arranged in corresponding pairs.

Exercise 54

Solve the following equations:

1 $x - y = 2$
   $x^2 + xy = 60$

2 $x + y = 7$
   $3x^2 + xy - y^2 = 81$

3 $2x + y = 5$
   $5x^2 - 3xy = 14$

4 $3x + y = 8$
   $x^2 - 5xy + 8y = 36$

5 $x + y + 1 = 0$
   $3x^2 - 5y^2 - 7 = 0$

6 $2x + 3y = 14$
   $4x^2 + 2xy + 3y^2 = 60$

7 $2x^2 - 5x + 4xy = 60$

   $3x - y = 9$

8 $2x + 3y = 5$

   $\dfrac{5}{y + 3} - \dfrac{9}{x} = 1$

9 $3x + y = 25$

   $xy = 28$

10 $2x - y = 8$

    $\dfrac{x^2}{9} + \dfrac{y^2}{16} = 5$

11 $\dfrac{1}{x} + \dfrac{1}{y} = \frac{3}{4}$

    $3x - y = 2$

148 Second type. Symmetric equations

Example 1: An example of this type is

$$x + y = 19 \tag{1}$$
$$xy = 84 \tag{2}$$

These two equations are called 'symmetric' because, if the letters x and y are interchanged throughout, the equations are unaltered. Other equations which are not strictly symmetric can be solved by the same method as these, and so are included.

In these particular equations number (1) is of the first degree, and so the previous method can be used; but this is an easy example to illustrate the special solution which can be employed with this type.

The aim of the method is to find the value of $x - y$. Then, since the value of $x + y$ is known, the rest of the solution is easy.

We are given

$$x + y = 19 \tag{1}$$
$$xy = 84 \tag{2}$$

Squaring both sides of (1)

$$x^2 + 2xy + y^2 = 361 \tag{3}$$

From (2)

$$4xy = 336 \qquad (4)$$

Subtracting (4) from (3) we get

$$x^2 - 2xy + y^2 = 25$$

$\therefore \qquad\qquad x - y = \pm 5.$

From (1) $\qquad\qquad x + y = 19.$

Adding $\qquad\qquad 2x = 19 \pm 5 = 24 \text{ or } 14.$

$\therefore \qquad\qquad x = 12 \text{ or } 7.$

Subtracting $\qquad 2y = 19 - (\pm 5)$

$\qquad\qquad\qquad\quad = 14 \text{ or } 24.$

$\therefore \qquad\qquad y = 7 \text{ or } 12.$

Or using $\qquad\qquad xy = 84$

(1) when $\qquad\qquad x = 12, y = 7,$

(2) when $\qquad\qquad x = 7, y = 12.$

The symmetry of the two equations appears in the solution.

Example 2: Solve the equations:

$$x^2 + y^2 = 89 \qquad (1)$$
$$xy = 40 \qquad (2)$$

Both equations are of the second degree. The aim in this case is to obtain both $x + y$ and $x - y$.

From (2)

$$2xy = 80 \qquad (3)$$

Adding (1) and (3)

$$x^2 + 2xy + y^2 = 169$$

$\therefore$

$$x + y = \pm 13.$$

Subtracting (3) from (1)

$$x^2 - 2xy + y^2 = 9.$$

$\therefore \qquad\qquad x - y = \pm 3.$

Associating the equations:

$$x + y = \pm 13.$$
$$x - y = \pm 3.$$

Adding $\qquad\qquad 2x = \pm 13 \pm 3 = \pm 16 \text{ or } \pm 10.$

$\therefore \qquad\qquad x = \pm 8 \text{ or } \pm 5.$

To get corresponding values of x we may now substitute in

$$xy = 40.$$

∴ when
$$x = +8, -8, +5, -5$$
$$y = +5, -5, +8, -8.$$

∴ associating these the solution is:

$$x = +8, y = +5$$
$$x = -8, y = -5$$
$$x = +5, y = +8$$
$$x = -5, y = -8.$$

Note: In the above examples the sum and difference of two numbers, $x + y, x - y$, are used. The same method would be used if we were given such an example as $x + 3y$. We should then aim at getting $(x + 3y)^2$, $(x - 3y)^2$, and ultimately $x - 3y$. There are many other variations in this type of solution. In some, such as

$$x^2 + y^2 = a \qquad (1)$$
$$x + y = b \qquad (2)$$

xy is not given, but it can be obtained by squaring equation (2) and subtracting equation (1).

Exercise 55

Solve the following equations:

1 $x + y = 8$
  $xy = 15$

2 $x + y = 9$
  $3xy - 42 = 0$

3 $x - y = 5$
  $xy = 24$

4 $x^2 + y^2 = 17$
  $x + y = 5$

5 $x - 2 = 2$
  $xy = 12$

6 $x^2 + y^2 = 17$
  $xy = 4$

7 $x^2 - xy + y^2 = 67$
  $xy = 18$

8 $x^2 - xy + y^2 = 43$
  $x + y = 8$

9 $x^2 + y^2 = 45$
  $x^2 - xy + y^2 = 27$

10 $2x + 3y = 9$
  $xy = 3$

149 Solving quadratic inequalities

As we saw in chapter 16, plotting a graph of the associated quadratic function illustrated one way in which the solution of a quadratic inequality could be obtained. However, as with the solution of quadratic equations, this method can be rather cumbersome. We will now look at another method using factorisation.

Example 1: Solve the quadratic inequality $x^2 - 6x + 8 < 0$

The left hand side of the inequality can be factorised as $(x - 2)(x - 4)$. So we now want to find the values of x for which $(x - 2)(x - 4) < 0$.

We want the product of the two brackets to be a negative number i.e. < 0. This can only happen if one of the brackets is a positive number and the other is a negative number. So we want $(x - 2) > 0$ and $(x - 4) < 0$ or $(x - 2) < 0$ and $(x - 4) > 0$.

The best way to handle these pairs of simultaneous inequalities is to use the number line method as we did in Section 69.

On the diagram we show that $(x - 2)$ is positive when $x > 2$ and negative when $x < 2$ by using the $+$ and $-$ signs. In the same way $(x - 4)$ is positive when $x > 4$ and negative when $x < 4$. For $(x - 2)(x - 4)$ to be less than 0 we need the values of x where we have one positive and one negative bracket. This only happens for values of x between 2 and 4, so the solution of $x^2 - 6x + 8 < 0$ is where $2 < x < 4$.

If we had wanted to find where $x^2 - 6x + 8 > 0$ we would have been looking for the values of x where the product of $(x - 2)$ and $(x - 4)$ was positive. This occurs *either* when both brackets are positive or when both brackets are negative. This happens when $x > 4$ or when $x < 2$. So the solution of $x^2 - 6x + 8 > 0$ is where $x < 2$ or where $x > 4$.

The advantage of this method is that it can be extended to cater for more complex situations where several brackets are involved. The next example shows how a cubic inequality can be solved.

Example 2: Solve the inequality $(2x - 1)(x + 2)(3 - x) > 0$

We want the product of the three brackets to be a positive number. This occurs *either* if all three of the brackets are positive *or* if one bracket is a positive number and the other two are negative numbers.

So on the number line diagram we want three +'s or one + and two −'s.

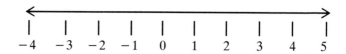

All three brackets are positive for values of x which are between 0.5 and 3. We have one positive and two negative brackets for values of x less than −2. So the solution of the inequality is where $0.5 < x < 3$ or where $x < -2$.

If the inequality had been $(2x - 1)(x + 2)(3 - x) \geq 0$ it would have been necessary to include the values −2, 0.5 and 3, i.e. $x \leq -2$ or $0.5 \leq x \leq 3$.

Exercise 56

1 Use the number line method to solve the inequality:

    (*a*) $(x - 2)(x - 3) > 0$ (*b*) $(x + 2)(x - 4) < 0$
    (*c*) $(x + 2)(x + 3) \geq 0$

2 Factorise the quadratic expression and then use the number line method to solve the inequality:

    (*a*) $x^2 - 3x + 2 > 0$ (*b*) $x^2 - 6x + 5 > 0$
    (*c*) $x^2 - 3x - 4 \geq 0$

3 Use the number line method to solve the inequality:

    (*a*) $(2x - 2)(3 - x) > 0$ (*b*) $(4x - 1)(x + 2) < 0$
    (*c*) $(2x + 1)(3x - 1) \geq 0$

4 Use the number line method to solve the inequality:

    (*a*) $(x - 2)(x - 3) > 0$ (*b*) $(2x + 1)(4x - 1)(x - 3) < 0$

5 Solve the inequality:

    (*a*) $x^2 - 3x < 0$ (*b*) $5x - x^2 \geq 0$
    (*c*) $(x - 2)(x - 2)(x - 5) > 0$

18

Indices

150 The meaning of an index

In Section 20 it was shown that the product of a number of equal factors such as $a \times a \times a \times a$ can be written in the form a^4, in which 4 is an index which indicates the number of factors.

Generalising this, if there be n such equal factors, where n is a positive integer, then a^n will be defined as follows:

$$a^n = a \times a \times \ldots \text{ to n factors}$$
and a^n is called the nth power of a.

151 Laws of indices

It was also shown in Sections 21, 22, 23 that when operations such as multiplication and division of powers of a number are performed, the laws which govern these operations can be deduced from the definitions of a power and an index given in Section 150.

General proofs of these laws will now be given.

I First law of indices: the law of multiplication

The special cases given in Section 21 lead to the general law as follows:

Let m and n be any positive integers.
By definition

$$a^m = a \times a \times a \times \ldots \text{ to m factors}$$
$$\text{and } a^n = a \times a \times a \times \ldots \text{ to n factors.}$$

Then $a^m \times a^n$ = (a × a × a × . . . to m factors) × (a × a × a × . . . to n factors).

But when two groups of factors are multiplied the factors in the groups are associated as one group of factors to give the product (see Section 21).

$\therefore$ $a^m \times a^n$ = a × a × a × a × . . . to (m + n) factors
$\qquad\qquad$ = a^{m+n} (by definition).

$\therefore$ the first law of indices is

$$a^m \times a^n = a^{m+n}$$

The law is clearly true when the product involves more than two powers. Thus $a^m \times a^n \times a^p = a^{m+n+p}$.

II   Second law of indices: the law of division

To find the value of $a^m \div a^n$.

With the same definition as before, and proceeding as in the special cases of Section 23.

$$a^m \div a^n = \frac{a \times a \times a \times \ldots \text{to m factors}}{a \times a \times a \times \ldots \text{to n factors}}.$$

After cancelling the n factors of the denominator with n corresponding factors in the numerator, there are left in the numerator m − n factors.

$$\therefore \ a^m \div a^n = a^{m-n}$$

Note: This proof assumes that m is greater than n. If m is less than n, there are n − m factors left in the denominator.

$$\therefore \text{ if } n > m \qquad a^m \div a^n = \frac{1}{a^{n-m}}.$$

This case will be examined later.

III   Third law of indices: the law of powers

By the definition of Section 150 such an expression as $(a^4)^3$ means the third power of a^4, or

$$\begin{aligned}
(a^4)^3 &= a^4 \times a^4 \times a^4 \\
&= a^{4+4+4} \text{ (by first index law)} \\
&= a^{4 \times 3} \\
&= a^{12}.
\end{aligned}$$

In general, if m and n are any positive integers by definition

$$(a^m)^n = a^m \times a^m \times a^m \ldots \text{ to n factors}$$
$$= a^{m+m+m} \ldots \text{ to n terms (first index law)}$$
$$= a^{m \times n}.$$

∴ the law of powers is

$$(a^m)^n = a^{m \times n}.$$

For powers of a product such as $(ab)^n$, see Section 22. In a similar way it may be shown that

$$(ab)^n = a^n \times b^n.$$

Exercise 57

Revision Exercise in Indices

Write down the values of the following:

1 $2a^4 \times 3a^5$

2 $a^6 \times a^6$

3 $\frac{1}{2}a^5 \times \frac{1}{4}a^3$

4 $2 \times 2^2 \times 2^3 \times 2^4$

5 $4 \times 4^2 \times 4^3$

6 $3a^3b^4 \times 2a^5b^3$

7 $x^4y^5z \times x^2y^7z^3$

8 $x^{m+1} \times x^{m-1}$

9 $a^{m+n} \times a^{m-n}$

10 $a^pb^q \times a^{p+q} \times b^{p-q}$

11 $a^{n+3} \times a^{n-3}$

12 $a^m \times a^n \times a^4$

13 $a^8 \div a^5$

14 $5x^6 \div 10x^3$

15 $36a^{10} \div 12a^5$

16 $2^{10} \div 2^4$

17 $3^5 \div 3^2$

18 $(-x^7) \div (-x^2)$

19 $15a^4b^2 \div -3ab$

20 $a^{2p} \div a^p$

21 $12x^{2p}y^{2q} \div 3x^py^q$

22 $\dfrac{a^{10} \times a^4}{a^7}$

23 $\dfrac{x^8}{x^4} \times \dfrac{x^3}{x^6}$

24 $a^{2n} \div a^{n-1}$

25 $a^{m+n} \div a^{m-n}$

26 $a^{n+4} \div a^{n-2}$

27 $\dfrac{4a^3}{3b^2} \div \dfrac{6a^2}{4b^3}$

28 $\dfrac{a}{b} \div \dfrac{b}{a}$

29 $(2^3)^2$

30 $(3^2)^3$

31 $(x^5)^2$

32 $(x^2)^5$

33 $(a^4)^4$

34 $(2a^2b^3)^2$

35 $\left(\dfrac{x^4y^5}{2y^2}\right)^3$

36 $(a^p)^3$

37 $(x^4)^n$ **38** $(3a^{2p})^3$

39 $\sqrt{a^4}$ **40** $\sqrt{x^8}$

41 $\sqrt{x^{16}}$ **42** $\sqrt{9a^6}$

43 $\sqrt{a^4b^2}$ **44** $\sqrt{\dfrac{x^6}{y^4}}$

45 $\sqrt{\dfrac{9a^8}{4b^6}}$ **46** $\sqrt{x^{2n}}$

---

152 Extension of the meaning of an index

It has so far been assumed that all indices are positive integers. The definition of a^n – viz.:

$$a^n = a \times a \times a \times \ldots \text{ to n factors}$$

is unintelligible except upon the assumption that n is a positive integer.

But algebra generalises, and we must therefore consider the possibility of attaching a meaning to an index in all cases.

153 Graph of 2^x

As a first step let us choose a suitable number, say 2, and plot some of the powers of it – i.e., draw the graph of 2^x, in which x represents any index. Calculating the values of these powers for some of the smaller integral values of x, we get a table of values as follows:

x	1	2	3	4
$y = 2^x$	2	4	8	16

When these points are plotted they appear to be on a smooth curve, as drawn in Fig. 40.

If we are justified in assuming that this curve is continuous and such that all the points on it satisfy the equation $y = 2^x$, then it follows that if any point be taken on the curve between the plotted points, its co-ordinates must also satisfy the law.

Therefore if any point, A, be taken on the curve at which $x = 1.5$ and $y = 2.8$, then it follows that $2^{1.5} = 2.8$.

Again at B, where x = 3.2 and y = 9, $2^{3.2} = 9$.

Hence, if the assumption is correct, that the curve is a continuous one, and the co-ordinates of any point on it satisfy $y = 2^x$, then it may be concluded that any number within the limits plotted can be expressed as a power of 2, and conversely any number can be used as the index of some power of 2.

A similar curve could be drawn exhibiting the powers of any other number – say, 10 – i.e., we could draw the curve of

$$y = 10^x.$$

Thus we could express any number as a power of 10.

Fig. 40.

Reverting to the curve PQ (Fig. 40), since it was plotted by taking values of x from unity, the curve began at P. If it be produced towards the y axis in the way it seems to curve, it will apparently cut the y axis at the point where y = 1.

This suggests that:

(1) The value of 2^x when x = 0 is 1 – i.e., $2^0 = 1$.
(2) The portion of the curve thus drawn should show values of 2^x between x = 0 and x = 1.

For example, when $x = \frac{1}{2}$, 2^x is approximately 1.4 – i.e., $2^{\frac{1}{2}} = 1.4$ approximately.

Again, the curve evidently does not end at the y axis, but can be further produced. This portion will correspond to negative values of x. Hence we infer that values of 2^x can be found when the index is negative.

154 Algebraical consideration of the extension of the meaning of indices

From the graph of powers it may be inferred that powers of a number exist whether the index be integral or fractional, positive or negative. We must now consider how such indices can be interpreted algebraically.

In seeking to find meanings for the new forms of indices we must be guided by one fundamental principle – viz.:

Every index must obey the laws of indices as discovered for positive integral numbers.

Otherwise it cannot be considered as an index.

155 Fractional indices

We will begin with the simple case of $a^{\frac{1}{2}}$.

To find a meaning for $a^{\frac{1}{2}}$.
Since $a^{\frac{1}{2}}$ must conform to the laws of indices
∴ by the first law

$$a^{\frac{1}{2}} \times a^{\frac{1}{2}} = a^{\frac{1}{2} + \frac{1}{2}}$$
$$= a^1 \text{ or } a.$$

∴ $a^{\frac{1}{2}}$ must be such a number that on being multiplied by itself the product is a.

But such a number is defined arithmetically as the square root of a.

$\therefore$ $\qquad\qquad$ $a^{\frac{1}{2}}$ must be defined as $\sqrt{a}$.

As an example $2^{\frac{1}{2}} = \sqrt{2} = 1.4134$ approx.

This agrees with the value of 1.4 found from the graph (Fig. 40). The reasoning above clearly holds in all such cases, and so we may infer that in general, if n be any positive integer

$$a^{\frac{1}{n}} = \sqrt[n]{a}.$$

To find a meaning for $a^{\frac{2}{3}}$.

Applying the first law of indices:

$$a^{\frac{2}{3}} \times a^{\frac{2}{3}} \times a^{\frac{2}{3}} = a^{\frac{2}{3} + \frac{2}{3} + \frac{2}{3}}$$
$$= a^2.$$
$\therefore$ $\qquad\qquad\qquad$ $a^{\frac{2}{3}} = \sqrt[3]{a^2}.$

Applying the reason generally, we can deduce that if m and n are any positive integers

$$a^{\frac{m}{n}} = \sqrt[n]{a^m}.$$

Thus $\qquad\qquad\qquad$ $a^{\frac{3}{4}} = \sqrt[4]{a^3}.$
$$a^{\frac{2}{5}} = \sqrt[5]{a^2}.$$

Indices which are in decimal form can be changed to vulgar fractions. Thus:

$$a^{0.25} = a^{\frac{1}{4}}$$
$$= \sqrt[4]{a}.$$

156 To find a meaning for a^0

Using the second law of indices, and n being any number,

$$a^n \div a^n = a^{n-n}$$
$$= a^0.$$
But $\qquad\qquad\qquad$ $a^n \div a^n = 1.$
$\therefore$ $\qquad\qquad\qquad\qquad$ $a^0 = 1.$

This confirms the conclusion reached in Section 153 from the graph. It should be noted that a represents any finite number. Therefore $a^0 = 1$, whatever the value of a.

Graphically, if curves of a^x are drawn for various values of a, as that of 2^x was drawn in Fig. 40, all these curves will pass through the point on the y axis which is unit distance from 0.

157   Negative indices

In considering the curve of 2^x in Fig. 40, the conclusion was reached that the curve could be continued for negative values of x. We must now find what meaning can be given algebraically to a negative index.

Consider a^{-1}.

This must obey the laws of indices.

∴ by the first law

$$a^{-1} \times a^{+1} = a^{-1+1}$$
$$= a^0$$
$$= 1.$$

Dividing by a^{+1} we get

$$a^{-1} = \frac{1}{a},$$

i.e., by a^{-1} we must mean 'the reciprocal of a'.

With the same reasoning

$$a^{-2} = \frac{1}{a^2}$$
$$a^{-3} = \frac{1}{a^3},$$

and in general

$$a^{-n} \times a^{+n} = a^{-n+n}$$
$$= a^0$$
$$= 1.$$
∴ $$a^{-n} = \frac{1}{a^n}.$$

∴ a^{-n} is defined as the reciprocal of a^n.

The following examples should be noted:

$$2a^{-3} = \frac{2}{a^3} \text{ (negative index applies to a only)}$$
$$a^{-\frac{1}{2}} = \frac{1}{\sqrt{a}}$$
$$\frac{1}{a^{-1}} = a$$

and generally $\dfrac{1}{a^{-n}} = a^n.$

It should be noted as a working rule that when a power of a number is transferred from the numerator of a fraction to the denominator, or vice-versa, the sign of the index is changed.

If n is a positive number a^n increases as n increases.

$\therefore a^{-n}$ or $\dfrac{1}{a^n}$ decreases as n increases.

Consequently the curve of 2^x will approach closer to the x axis for negative values of x.

This agrees with the course which the curve in Fig. 40 appeared to be taking.

158   Standard forms of numbers

Indices may be usefully employed in what are called standard forms, by means of which we can express clearly and concisely certain numbers which are very large or very small.

The method of formation of them will be seen from the following examples:

$$26 = 2.6 \times 10^1$$
$$260 = 2.6 \times 10^2$$
$$2600 = 2.6 \times 10^3$$
$$26\,000 = 2.6 \times 10^4$$
$$260\,000 = 2.6 \times 10^5 \text{ and so on.}$$

When a number is written in standard form, one digit only is retained in the whole number part, and this is multiplied by a power of 10 to make it equal to the number. It should be noted that the index of the power of 10 is equal to the number of figures which follow the figure retained in the integral part of the number.

Thus in $547\,000\,000$, when written in standard form, 5 only is retained in the whole number part and 8 figures follow it.

$$\therefore \qquad 547\,000\,000 = 5.47 \times 10^8.$$

Numbers which are less than unity may be similarly changed into standard form by using negative indices.

Thus $0.26 \quad = 2.6 \div 10 \text{ or } 2.6 \times 10^{-1}$
$0.026 \quad = 2.6 \div 10^2 \text{ or } 2.6 \times 10^{-2}$
$0.0026 \quad = 2.6 \div 10^3 \text{ or } 2.6 \times 10^{-3}$
$0.000\,26 = 2.6 \div 10^4 \text{ or } 2.6 \times 10^{-4}, \text{ etc.}$

It should be noted that the numerical part of the index is one more than the number of zeros following the decimal point in the original number.

159   Operations with standard forms

In these operations the rules of indices must be observed.

Examples

(1) $(1.2 \times 10^4) \times (2.3 \times 10^3) = (1.2 \times 2.3) \times (10^4 \times 10^3)$
$= 2.76 \times 10^7$.

(2) $(4.8 \times 10^8) \div (1.6 \times 10^{-3}) = (4.8 \div 1.6) \times (10^8 \div 10^{-3})$
$= 3 \times 10^{11}$.

Exercise 58

Note: When necessary in the following examples take $\sqrt{2} = 1.414$,
$\sqrt{3} = 1.732$, $\sqrt{5} = 2.236$, $\sqrt{10} = 3.162$.

1 State what meanings can be given to the following:

(1) $3^{\frac{1}{4}}$ (2) $8^{\frac{1}{3}}$ *wRoNbr*   (3) $a^{\frac{1}{3}}$
(4) $2^{0.5}$ (5) $3^{0.5}$ $\sqrt{1}$ (6) $a^{0.75}$

2 State what meanings can be given to the following:

(1) $10^{\frac{3}{4}}$ (2) $4^{\frac{2}{3}}$ *wRoNbr* (3) $a^{\frac{3}{4}}$
(4) $10^{1.5}$ (5) $2^{2.5}$ (6) $10^{1.25}$

3 Write down as simply as possible the meaning of:

(1) $4^{\frac{3}{2}}$ (2) $8^{\frac{3}{4}}$ (3) $a^{1.75}$
(4) $a^{\frac{7}{3}}$ (5) $a^{1.2}$ (6) $a^{0.2}$

4 Find the numerical values of:

(1) $4^{1.5}$ (2) $16^{1.25}$ (3) $9^{1.5}$
(4) $100^{\frac{3}{2}}$ (5) $10^{\frac{3}{2}}$ (6) $(\frac{1}{4})^{0.5}$

5 Find the values of 3^{-2}, 3^{-1}, 3^0, 3^1, $3^{1.5}$, 3^2, $3^{2.5}$. Plot these and draw the curve which contains the points.

6 Find the values of:

(1) $2^2 \times 2^{\frac{1}{2}}$ (2) $3 \times 3^{\frac{1}{2}} \times 3^{\frac{3}{2}}$
(3) $10^{\frac{1}{2}} \div 10^{\frac{3}{2}}$ (4) $a^{\frac{3}{4}} \times a^{\frac{1}{4}}$
(5) $2^{\frac{5}{2}} \div 2^{\frac{3}{2}}$ (6) $a^{1.5} \div a^{0.3}$

7 Write down the meanings with positive indices of:

$$4^{-1}, 3a^{-2}, 2^{-\frac{1}{4}}, \frac{1}{3^{-1}}, \frac{3}{a^{-2}}, 10^{-3}.$$

8 Write with positive indices:

(1) x^{-3} (2) $x^{-\frac{1}{2}}$ (3) $\dfrac{1}{x^{-3}}$

(4) $(x^{-3})^2$ (5) $\left(\dfrac{1}{x}\right)^{-1}$ (6) $\dfrac{1}{2x^{-\frac{1}{2}}}$

9 Find the values of:

(1) $8^{\frac{2}{3}}$ (2) $25^{\frac{3}{2}}$ (3) $(10^2)^{\frac{3}{2}}$

(4) $(5^{-3})^2$ (5) $\dfrac{2}{2^{-3}}$ (6) $81^{\frac{1}{4}}$

10 Find the values of:

(1) $(\frac{1}{2})^{-2}$ (2) $(\frac{2}{3})^{-3}$ (3) $16^{-0.5}$

(4) $(36)^{-0.5}$ (5) $(4)^{1.5}$ (6) $(\frac{1}{4})^{2.5}$

11 Find the value of $a^4 \times a^{-2} \times a^{\frac{1}{2}}$ when $a = 2$.

12 Write down the simplest forms of:

(1) $a^{\frac{1}{2}} \times a^{\frac{1}{3}}$ (2) $10^3 \times 10^{-\frac{1}{2}}$

(3) $a^3 \div (-a)^6$ (4) $-a^3 \div (-a)^5$

(5) $x^{2n} \div x^{-n}$ (6) $x^n \div x^{-n}$

13 Find the values of:

(1) $a^3 \times a^0$ (2) $a^0 \times 1$ (3) $a^0 \times 0$

14 Write down in 'standard form' the answers to the following:

(1) $(2.2 \times 10^5) \times (1.6 \times 10^4)$

(2) $(7.1 \times 10^3) \times (2.3 \times 10^3)$

(3) $(4.62 \times 10^5) \div (2.1 \times 10^3)$

(4) $(7.4 \times 10^6) \times (5 \times 10^{-4})$

(5) $(1.2 \times 10^{-3}) \times (2.1 \times 10^{-4})$

19

Logarithms

160 A system of indices

In the previous chapter it was seen that by means of the graph of 2^x it is possible, within the limits of the graph, to express any number as a power of 2. This was confirmed algebraically.

For every number marked on the y axis and indicated on the graph, there is a corresponding index which can be read on the x axis. These constitute a system of indices by which numbers can be expressed as powers of a common basic number 2.

Similarly, by drawing graphs such as 3^x, 5^x, 10^x, numbers can be expressed as powers of 3, 5, or 10, or any other basic number.

Thus in all such cases it is possible to formulate systems of indices which, for any number A, would enable us to determine what power that number is of any other number B, which is called the base of the system.

This possibility of expressing any number as a power of any other number, and thus of the formation of a system of indices, as stated above, leads to practical results of great importance. It enables us to carry out, easily and accurately, calculations which would otherwise be laborious or even impossible. The fundamental ideas underlying this can be illustrated by means of a graph of powers similar to that drawn in Fig. 40. For this purpose we will use 10 as the base of the system and draw the graph of $y = 10^x$.

As powers of 10 increase rapidly, it will be possible to employ only small values for x, if the curve is to be of any use for our purpose. To obtain those powers we must use the rules for indices which were formulated in the previous chapter.

From arithmetic we know that $\sqrt{10} = 3.16$ app., i.e., $10^{\frac{1}{2}} = 3.16$.

Then

$$10^{0.25} = 10^{\frac{1}{4}} = (10^{\frac{1}{2}})^{\frac{1}{2}} = (3.16)^{\frac{1}{2}} = 1.78 \text{ app. (by arithmetic).}$$
$$10^{0.75} = 10^{\frac{3}{4}} = 10^{\frac{1}{2} + \frac{1}{4}} = 10^{\frac{1}{2}} \times 10^{\frac{1}{4}} = 3.16 \times 1.78 = 5.62 \text{ app.}$$
$$10^{0.125} = 10^{\frac{1}{8}} = (10^{\frac{1}{4}})^{\frac{1}{2}} = (1.78)^{\frac{1}{2}} = 1.33 \text{ app.}$$

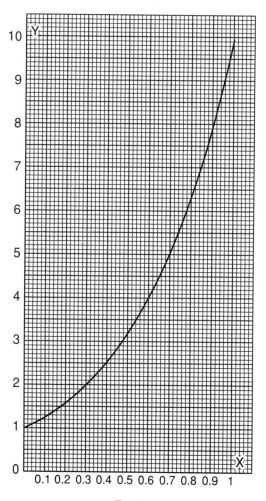

Fig. 41.

In this way a table of values for the curve can be compiled as follows:

x	0	0.125	0.25	0.5	0.75	0.875	1
10^x	1	1.33	1.78	3.16	5.62	7.5	10

The resulting curve is shown in Fig. 41. The following examples illustrate the use that can be made of it in calculations.

Example 1: Find from the graph the value of 1.8×2.6.

From the graph

$$1.8 = 10^{0.26}$$
$$2.6 = 10^{0.42}$$
$$\therefore \quad 1.8 \times 2.6 = 10^{0.26} \times 10^{0.42}$$
$$= 10^{0.26 + 0.42} \text{ (first law of indices)}$$
$$= 10^{0.68}$$
$$= 4.6 \text{ from the graph.}$$

Example 2: Find $\sqrt[3]{9}$.

From the graph

$$9 = 10^{0.96}$$
$$\sqrt[3]{9} = 9^{\frac{1}{3}}$$
$$= (10^{0.96})^{\frac{1}{3}}$$
$$= 10^{0.32} \text{ (third law of indices)}$$
$$= 2.1 \text{ from the graph.}$$

161   A system of logarithms

Although interesting as illustrating the principles involved, the above has little practical value for purposes of calculation, since we depend upon the readings from a curve which is necessarily limited in size and is not sufficiently accurate.

For practical purposes it is obvious that tables of the indices used, calculated to a suitable degree of accuracy, are necessary. Such tables have been compiled and are available for the purpose. For their compilation a more advanced knowledge of mathematics is required than it included in this volume.

The tables are constructed with 10 as a suitable base. They give the indices which indicate for all numbers, within the scope of the table, the powers they are of 10.

Such a table is called a system of logarithms, and the number 10, with respect to which the logarithms are calculated, is called the base of the system.

A logarithm to base 10 may be defined as follows:

The logarithm of a number to base 10 is the index of the power to which 10 must be raised to produce the number.

162 Notation for logarithms

You may wonder why another, unfamiliar term is employed as a name for an index. One reason for this will be seen from the following:

Let n be any positive number.
Let x be its index to base 10.
Then n = 10^x.

This is in reality a formula. If it is required to 'change the subject of the formula' (see Section 57) and express x in terms of the other letters, there is a difficulty in doing this concisely. Using words we could write:

x = index of power of n to base 10.

This is cumbersome, so we employ the word 'logarithm',[1] abbreviated to 'log' as follows:

$$x = \log_{10} n$$

the number indicating the base being inserted as shown.

If the base is e, we write x = $\log_e$ n.

In this form x is expressed as a function of n, whereas in the form n = 10^x, n is expressed as a function of x.

You should be able to change readily from one form to another.

Examples

(1) We saw in Section 160 that 56.2 = $10^{1.75}$.
  In log form this is   1.75 = $\log_{10}$ 56.2.

(2)   1024 = 2^{10}.
  ∴ $\log_2$ 1024 = 10.

---

[1] The choice of the word logarithm can be explained only by the history of the word. You could consult *A Short History of Mathematics*, by W. W. R. Ball.

(3) $1000 = 10^3$.

∴ $\log_{10} = 3$.

(4) $81 = 3^4$.

∴ $\log_3 81 = 4$.

For ordinary calculations 10 is the most suitable base for a system of logarithms, but in more advanced mathematics a different base is required (see Section 170).

163   Characteristic of a logarithm

The integral or whole number part of a logarithm is called the characteristic. This can always be determined by inspection when logarithms are calculated to base 10, as will be seen from the following considerations:

Since $10^0 = 1$, $\log_{10} 1$ $= 0$
                $10^1 = 10$, $\log_{10} 10$ $= 1$
                $10^2 = 100$, $\log_{10} 100$ $= 2$
                $10^3 = 1000$, $\log_{10} 1000$ $= 3$
                $10^4 = 10\,000$, $\log_{10} 10\,000$ $= 4$

and so on.
From these results we see that,

for numbers between   1 and 10   the characteristic is 0
       "   10   100   "   1
       "   100   1000   "   2
       "   1000   10 000   "   3

and so on.

It is evident that the characteristic is always one less than the number of digits in the whole number part of the number.

Thus in $\log_{10} 3758.7$ the characteristic is 3
             $\log_{10} 375.87$ "   2
             $\log_{10} 37.587$ "   1.

Thus the characteristics may always be determined by inspection, and consequently are not given in the tables. This is one advantage of having 10 for a base.

164   Mantissa of a logarithm

The decimal part of a logarithm is called the mantissa.

In general the mantissa can be calculated to any required

number of figures, by the use of higher mathematics. In most tables, such as those given in this volume, the mantissa is calculated to four places of decimals approximately. In Chambers' *Book of Tables* they are calculated to seven places of decimals.

The mantissa alone is given in the tables, and the following example will show the reason why:

$$\log_{10} 168.3 = 2.2261.$$
$$\therefore \qquad 168.3 = 10^{2.2261}.$$
$$\therefore \qquad 168.3 \div 10 = 10^{2.2261} \div 10^{1}.$$
$$= 10^{1.2261}.$$
$$\therefore \qquad \log_{10} 16.83 = 1.2261.$$

Similarly $\qquad \log_{10} 1.683 = 0.2261$

and $\qquad \log_{10} 1683 = 3.2261.$

Thus, if a number is multiplied or divided by a power of 10, the characteristic of the logarithm of the result is changed, but the mantissa remains unaltered. This may be expressed as follows:

Numbers having the same set of significant figures have the same mantissa in their logarithms.

165 To read a table of logarithms

With the use of the above rules relating to the characteristic and mantissa of logarithms, the student should have no difficulty in reading a table of logarithms.

Below is a portion of such a table, giving the logarithms of numbers between 33 and 37.

No.	Log.	1	2	3	4	5	6	7	8	9	1	2	3	4	5	6	7	8	9
31	4914	4928	4942	4955	4969	4983	4997	5011	5024	5038	1	3	4	6	7	8	10	11	12
32	5051	5065	5079	5092	5105	5119	5132	5145	5159	5172	1	3	4	5	7	8	9	11	12
33	5185	5198	5211	5224	5237	5250	5263	5276	5289	5302	1	3	4	5	6	8	9	10	12
34	5315	5328	5340	5353	5366	5378	5391	5403	5416	5428	1	3	4	5	6	8	9	10	11
35	5441	5453	5465	5478	5490	5502	5514	5527	5539	5551	1	2	4	5	6	7	9	10	11

(1)   (2)

The figures in column 1 in the complete table are the numbers from 1 to 99. The corresponding number in column 2 is the mantissa of the logarithm. As previously stated, the characteristic is not given, but can be written down by inspection. Thus $\log_{10} 31 = 1.4914$, $\log_{10} 310 = 2.4914$, etc.

If the number has a third significant figure, the mantissa will be found in the appropriate column of the next nine columns.

Thus $\log_{10} 31.1 = 1.4928,$
                     $\log_{10} 31.2 = 1.4942,$ and so on.

If the number has a fourth significant figure space does not allow us to print the whole of the mantissa. But the next nine columns of what are called 'mean differences' give us for every fourth significant figure a number which must be added to the mantissa already found for the first three significant figures. Thus if we want $\log_{10} 31.67$, the mantissa for the first three significant figures 316 is 0.4997. For the fourth significant figure 7 we find in the appropriate column of mean differences the number 10. This is added to 0.4997 and so we obtain for the mantissa 5007.

∴ $\log_{10} 31.67 = 1.5007.$

Anti-logarithms

You are usually provided with a table of anti-logarithms which contains the numbers corresponding to given logarithms. These could be found from a table of logarithms but it is quicker and easier to use the anti-logarithms, which are given at the end of this book.

The tables are similar in their use to those for logarithms, but we must remember:

(1) That the mantissa of the log only is used in the table.
(2) When the significant figures of the number have been obtained, you must proceed to fix the decimal point in them by using the rules which we have considered for the characteristic.

Example: Find the number whose logarithm is 2.3714.

First using the mantissa – viz., 0.3714 – we find from the anti-logarithm table that the number corresponding is given as 2352. These are the first four significant figures of the number required.

Since the characteristic is 2, the number must lie between 100 and 1000 (see Section 163) and therefore it must have 3 significant figures in the integral part.

∴ the number is 235.2.

Note: As the log tables which will be usually employed by the beginner are all calculated to base 10, the base in further work will be omitted when writing down logarithms. Thus we shall write log 235.2 = 2.3714, the base 10 being understood.

Exercise 59

1 Write down the characteristics of the logarithms of the following numbers:

15, 1500, 31 672, 597, 8, 800 000
51.63, 3874.5, 2.615, 325.4

2 Read from the tables the logarithms of the following numbers:

(1) 5, 50, 500, 50 000
(2) 4.7, 470, 47 000
(3) 52.8, 5.28, 528
(4) 947.8, 9.478, 94 780
(5) 5.738, 96.42, 6972

3 Find, from the tables, the numbers of which the following are the logarithms:

(1) 2.65, 4.65, 1.65
(2) 1.943, 3.943, 0.943
(3) 0.6734, 2.6734, 5.6734
(4) 3.4196, 0.7184, 2.0568

---

166 Rules for the use of logarithms

In using logarithms for calculations we must be guided by the laws which govern operations with them. Since logarithms are indices, these laws must be the same in principle as those of indices. These laws are given below; formal proofs are omitted. They follow directly from the corresponding index laws.

(1) **Logarithm of a product**
    The logarithm of the product of two or more numbers is equal to the sum of the logarithms of these numbers (see first law of indices).
    Thus if p and q be any numbers

$$\log (p \times q) = \log p + \log q.$$

(2) **Logarithm of a quotient**
    The logarithm of p divided by q is equal to the logarithm of p diminished by the logarithm of q (see second law of indices).
    Thus $\log (p \div q) = \log p - \log q.$

(3) **Logarithm of a power**

The logarithm of a power of a number is equal to the logarithm of the number multiplied by the index of the power (see third law of indices).

Thus $$\log a^n = n \log a.$$

(4) **Logarithm of a root**

This is a special case of the above (3).

Thus $$\log \sqrt[n]{a} = \log a^{\frac{1}{n}}$$
$$= \frac{1}{n} \log a.$$

167 Examples of the use of logarithms

Example 1: Find the value of 57.86×4.385.

Let $\quad$ x = 57.86×4.385.

Then $\quad$ log x = log 57.86 + log 4.385
$\qquad$ = 1.7624 + 0.6420
$\qquad$ = 2.4044
$\qquad$ = log 253.7.
∴ $\qquad$ x = 253.7.

No.	log.
57.86	1.7624
4.385	0.6420
253.7	2.4044

Notes:

(1) You should remember that the logs in the tables are correct to four significant figures only. Consequently you cannot be sure of four significant figures in the answer. It would be more correct to give the above answer as 254, correct to three significant figures.

(2) You are advised to adopt some systematic way of arranging the actual operations with logarithms. Such a method is shown above.

Example 2: Find the value of

$$\frac{5.672 \times 18.94}{1.758}.$$

Let $\quad$ x = $\dfrac{5.672 \times 18.94}{1.758}$.

∴ log x = log 5.672 + log 18.94 − log 1.758
$\qquad$ = 0.7538 + 1.2774 − 0.2450
$\qquad$ = 1.7862
$\qquad$ = log 61.12.
∴ $\qquad$ x = 61.12

No.	log.
5.672	0.7538
18.94	1.2774
	2.0312
1.758	0.2450
61.12	1.7862

or $\qquad$ x = 61.1 (to three significant figures).

Example 3: Find the fifth root of 721.8.

Let $\qquad$ $x = \sqrt[5]{721.8}$
$\qquad\qquad = (721.8)^{\frac{1}{5}}$
Then $\qquad$ $\log x = \frac{1}{5} \log 721.8$ (see Section 166(4))
$\qquad\qquad\quad = \frac{1}{5} (2.8584)$
$\qquad\qquad\quad = 0.5717.$
∴ $\qquad\qquad x = 3.730.$

Exercise 60

Use logarithms to find the values of the following:

1	23.4×14.73	**14**	$(15.23)^2 \times 3.142$
2	43.97×6.284	**15**	$(5.98)^2 \div 16.47$
3	987.4×1.415	**16**	$\dfrac{(91.5)^2}{4.73 \times 16.92}$
4	$42.7 \times 9.746 \times 14.36$		
5	$28.63 \div 11.95$	**17**	$\dfrac{(8.97)^2 \times (1.059)^2}{57.7}$
6	$43.97 \div 6.284$		
7	$23.4 \div 14.73$	**18**	$\dfrac{4798}{(56.2)^2 \div (9.814)^3}$
8	$927.8 \div 4.165$		
9	$94.76 \times 4.195 \div 27.94$	**19**	$\sqrt[4]{3.417}$
10	$\dfrac{15.36 \times 9.47 \times 11.48}{5.632 \times 21.85}$	**20**	$\sqrt[3]{4.872}$
11	$(9.478)^3$	**21**	$\sqrt[3]{1.625^2 \times 4.738}$
12	$(51.47)^2$		
13	$(1.257)^5$	**22**	$\sqrt[5]{61.5 \times 2.73}$
23	If $\pi r^2 = 78.6$ find r when $\pi = 3.142$.		
24	If $\frac{4}{3}\pi r^3 = 15.5$, find r when $\pi = 3.142$.		

168 Logarithms of numbers between 0 and 1

In Section 163 we gave examples of powers of 10 when the index is a positive integer. We will now consider cases in which the indices are negative.

Thus $\quad$ $10^1 = 10$ $\qquad$ ∴ $\log_{10} 10 = 1$
$\qquad\quad$ $10^0 = 1$ $\qquad\quad$ ∴ $\log_{10} 1 = 0$
$\qquad\quad$ $10^{-1} = \frac{1}{10} = 0.1$ $\quad$ ∴ $\log_{10} 0.1 = -1$
$\qquad\quad$ $10^{-2} = \dfrac{1}{10^2} = 0.01$ $\quad$ ∴ $\log_{10} 0.01 = -2$
$\qquad\quad$ $10^{-3} = \dfrac{1}{10^3} = 0.001$ ∴ $\log_{10} 0.001 = -3$ $\qquad$ etc.

From these results we may deduce that:
The logarithms of numbers between 0 and 1 are always negative.

We have seen (Section 164) that if a number be divided by 10, we obtain the log of the result by subtracting 1.

Thus if
$$\log 49.8 \quad = 1.6972$$
$$\log\ \ 4.98 \quad = 0.6972$$
$$\log\ \ 0.498 \quad = 0.6972 - 1$$
$$\log\ \ 0.0498 \quad = 0.6972 - 2$$
$$\log\ \ 0.004\ 98 = 0.6972 - 3$$

From the above,
$$\log 0.498 = 0.6972 - 1$$
$$= -0.3028.$$

Now, in the logs of numbers greater than unity, the mantissa remains the same when the numbers are multiplied or divided by powers of 10 (see Section 163), i.e. with the same significant figures we have the same mantissa.

It would clearly be a great advantage if we could find a system which would enable us to use this rule for numbers less than unity, and so avoid, for example, having to write

$$\log 0.498 \text{ as } -0.3028.$$

This can be done by not carrying out the subtraction as shown above, and writing down the characteristic as negative. But to write log 0.498 as $0.6972 - 1$ would be awkward. Accordingly we adopt the notation $\bar{1}.6972$, writing the minus sign above the characteristic.

It is very important to remember that

$$\bar{1}.6972 = -1 + 0.6972.$$

Thus in logarithms written in this way the characteristic is negative and the mantissa is positive.

With this notation
$$\log 0.0498 \quad = \bar{2}.6972$$
$$\log 0.004\ 98 \quad = \bar{3}.6972$$
$$\log 0.000\ 498 = \bar{4}.6972 \quad \text{etc.}$$

Note: The student should note that the negative characteristic is numerically one more than the number of zeros after the decimal point.

Example 1: From the tables find the logs of 0.3185, 0.031 85 and 0.003 185.

Using the portion of the tables in Section 165, we see that the

mantissa for 0.3185 will be 0.5031.
 Also the characteristic is − 1.

∴   log 0.3185   = $\overline{1}$.5031.
Similarly   log 0.031 85   = $\overline{2}$.5031
and   log 0.003 185 = $\overline{3}$.5031.

Example 2: Find the number whose log is $\overline{3}$.5416.

From the From the anti-log tables we find that the significant figures of the number whose mantissa is 5416 are 3840. As the characteristic is − 3, there will be two zeros after the decimal point.

∴ the number is   0.003 480

(correct to 4 significant figures).

Exercise 61

1 Write down the logarithms of:

(1) 2.798, 0.2798, 0.027 98
(2) 4.264, 0.4264, 0.004 264
(3) 0.009 783, 0.000 978 3, 0.9783
(4) 0.064 51, 0.6451, 0.000 645 1

2 Write down the logarithms of:

(1) 0.059 86   (4) 0.000 092 75
(2) 0.000 473   (5) 0.5673
(3) 0.007 963   (6) 0.079 86

3 Find the numbers whose logarithms are:

(1) $\overline{1}$.3342   (4) $\overline{4}$.6437
(2) $\overline{3}$.8724   (5) $\overline{1}$.7738
(3) $\overline{2}$.4871   (6) $\overline{8}$.3948

169 Operations with logarithms which are negative

Care is needed in operating with the logarithms of numbers which lie between 0 and 1, since they are negative and, as shown above, are written with the characteristic negative and the mantissa positive.
 A few examples will show the method of working.

Example 1: Find the sum of the logarithms:

$$\overline{1}.6173, \overline{2}.3415, \overline{1}.6493, \overline{0}.7374.$$

Arranging thus

$$\begin{array}{r} \overline{1}.6173 \\ \overline{2}.3415 \\ \overline{1}.6493 \\ \overline{0}.7374 \\ \hline \overline{2}.3455 \end{array}$$

The point to be specially remembered is that the 2 which is carried forward from the addition of the mantissae is positive, since they are positive. Consequently the addition of the characteristics becomes

$$-1 - 2 - 1 + 0 + 2 = -2.$$

Example 2: From the logarithm $\overline{1}.6175$ subtract the log $\overline{3}.8463$.

$$\begin{array}{r} \overline{1}.6175 \\ \overline{3}.8463 \\ \hline 1.7712 \end{array}$$

Here in 'borrowing' to subtract the 8 from the 6, the -1 in the top line becomes -2, consequently on subtracting the characteristics we have

$$-2 - (-3) = -2 + 3 = +1.$$

Example 3: Multiply $\overline{2}.8763$ by 3.

$$\begin{array}{r} \overline{2}.8763 \\ 3 \\ \hline \overline{4}.6289 \end{array}$$

From the multiplication of the mantissa, 2 is carried forward. But this is positive and as $(-2) \times 3 = -6$, the characteristic becomes $-6 + 2 = -4$.

Example 4: Multiply $\overline{1}.8738$ by 1.3.

In a case of this kind it is better to multiply the characteristic and mantissa separately and add the results.

Thus
$$0.8738 \times 1.3 = 1.135\,94$$
$$-1 \times 1.3 = -1.3.$$

-1.3 is wholly negative and so we change it to $\overline{2}.7$, to make the mantissa positive.

Then the product is the sum of

$$1.135\ 94$$
$$\overline{2}.7$$
_____
$$\overline{1}.835\ 94$$
or   1.8359 approx.

Example 5: Divide $\overline{5}.3716$ by 3.

Here the difficulty is that on dividing $\overline{5}.3716$ by 3 there is a remainder 2 which is negative, and cannot therefore be carried on to the positive mantissa. To get over the difficulty we write:

$$-5 = -6 + 1$$
or the log as $-6 + 1.3716.$

Then the division of the -6 gives us -2 and the division of the positive part 1.3716 gives 0.4572, which is positive. Thus the complete quotient is $\overline{2}.4572$. The work might be arranged thus:

$$3)\overline{6} + 1.3716$$
_____
$$\overline{2} + 0.4572$$
_____
$$= \overline{2}.4572$$

Exercise 62

1 Add together the following logarithms:

(1) $\overline{2}.5178 + 1.9438 + 0.6138 + \overline{5}.5283$
(2) $3.2165 + \overline{3}.5189 + \overline{1}.3297 + \overline{2}.6475$

2 Find the values of:

(1) $4.2183 - 5.6257$ (3) $\overline{1}.6472 - \overline{1}.9875$
(2) $0.3987 - \overline{1}.5724$ (4) $\overline{2}.1085 - \overline{5}.6271$

3 Find the values of:

(1) $\overline{1}.8732 \times 2$ (4) $\overline{1}.5782 \times 1.5$
(2) $\overline{2}.9456 \times 3$ (5) $\overline{2}.9947 \times 0.8$
(3) $\overline{1}.5782 \times 5$ (6) $\overline{2}.7165 \times 2.5$

4 Find the values of:

(1) $\overline{3}.9778 \times 0.65$ (4) 2.1342×-0.4
(2) $\overline{2}.8947 \times 0.84$ (5) 1.3164×-1.5
(3) $\overline{1}.6257 \times 0.6$ (6) $\overline{1}.2976 \times -0.8$

5 Find the values of:

(1) $\overline{1}.4798 \div 2$

(2) $\overline{2}.5637 \div 5$

(3) $\overline{4}.3178 \div 3$

(4) $\overline{3}.1195 \div 2$

(5) $\overline{1}.6173 \div 1.4$

(6) $\overline{2}.3178 \div 0.8$

170　Change of base of a system of logarithms

Although logs calculated to base 10 are usually employed for calculations in more advanced mathematics, as well as in engineering, the logs which naturally arise are calculated to a base which is given by the series

$$1 + \frac{1}{1} + \frac{1}{1.2} + \frac{1}{1.2.3} + \frac{1}{1.2.3.4} + \ldots \text{ to infinity.}$$

This series is denoted by e, and its value can be calculated to any required degree of accuracy by taking sufficient terms. To 5 places of decimals, e = 2.718 28.

Logs calculated to this base are called Naperian logarithms, after Lord Napier, who discovered them in 1614, using this base. They are also called Natural logarithms or Hyperbolic logarithms.

Possessing only tables of logs to base 10, you may require to use the logs of numbers to base e, and must therefore know how to find them.

The relations between the logs of numbers to different bases is found as follows:

Let n be any number.

Let a and b be two bases.

Suppose that logs to base b are known, and we require to find them to base a.

Let　　　$\log_b n = y, \therefore n = b^y.$

Then　　$\log_a n = \log_a (b^y)$

　　　　　　　　$= y \log_a b.$　(Section 166, rule 3.)

$\therefore$　　　$\log_a n = \log_b n \times \log_a b.$

Thus, knowing the log of a number to a base b, we find its log to base a by multiplying, whatever the number, by $\log_a b$.

In the above result let b = 10 and a = e.

Then　　　　　　$\log_e n = \log_{10} n \times \log_e 10.$

In this result   let n = e

then $\log_e e = \log_{10} e \times \log_e 10$

       but $\log_e e = 1$.

$\therefore$ $\log_{10} e \times \log_e 10 = 1$.

$\therefore$ $\log_e 10 = \dfrac{1}{\log_{10} e}$.

$\therefore$ in the rule $\log_e n = \log_{10} n \times \log_e 10$

we can write $\log_e n = \log_{10} n \times \dfrac{1}{\log_{10} e}$.

Thus both logs on the right-hand side are to base 10.

Now $\log_{10} e = 0.4343$

and $\log_{10} e = \dfrac{1}{0.4343} = 2.3026$.

Hence to change from base 10 to base e, we may use either of the following:

       (1) $\log_e n = \log_{10} n \times 2.3026$

or   (2) $\log_e n = \log_{10} n \div 0.4343$.

Example: Find $\log_e 50$.

Using $\log_e 50 = \log_{10} 50 \times 2.3026$

we have $\log_e 50 = 1.6990 \times 2.3026$.

Evaluating the right-hand side by use of logs we get

       $\log_e 50 = 3.913$.

Summary of the laws relating to logarithms, together with some special points the truth of which will be obvious.

(1) The logarithm of the base itself is always unity.
(2) The logarithm of 1 is always zero, whatever the base.
(3) The logarithms of all numbers less than unity are negative.
(4) The logarithm of a number is equal to $-$(the log of its reciprocal).

Thus $\log_a n = -\log_a \dfrac{1}{n}$.

(5) $\log_a (x \times y) = \log_a x + \log_a y$.
(6) $\log_a (x \div y) = \log_a x - \log_a y$.
(7) $\log_a x^n = n \log_a x$.
(8) $\log_a \sqrt[n]{x} = \dfrac{1}{n} \log_a x$.

Exercise 63

Miscellaneous Exercises in the Use of Logarithms

1 15.62×0.987

2 0.4732×0.694

3 0.513×0.0298

4 $75.94 \times 0.0916 \times 0.8194$

5 $9.463 \div 15.47$

6 $0.9635 \div 29.74$

7 $27.91 \div 569.4$

8 $0.0917 \div 0.5732$

9 $5.672 \times 14.83 \div 0.9873$

10 $(0.9173)^2$

11 $(0.4967)^3$

12 $\sqrt[3]{1.715}$

13 $\sqrt[5]{647.2} \div (3.715)^3$

14 $\frac{1}{2}(48.62)^{\frac{1}{3}}$

15 $\sqrt[3]{\dfrac{9.728}{3.142}}$

16 $(1.697)^{2.4}$

17 $(19.72)^{0.57}$

18 $(0.478)^{3.1}$

19 $(5.684)^{-1.12}$

20 $(0.5173)^{-3.4}$

21 $\sqrt[4]{0.016\ 97}$

22 $(0.1478)^2 \div 0.6982$

23 $\sqrt[3]{0.8172} \div \sqrt[3]{0.7658}$

24 $9.74^2 - 5.66^2$. (*Hint*: This should first be factorized and changed to a product.)

25 $\dfrac{9.32 \times 0.761}{\sqrt{18.2}}$

26 $\sqrt[3]{\dfrac{647.3 \times 3.2}{3.142 \times 10.78}}$

27 $\sqrt{(3.62)^2 + (5.47)^2 + (6.91)^2}$

28 Find the value of πr^2 when $\pi = 3.142$ and $r = 16.89$.

29 Find the value of $\frac{4}{3}\pi r^3$ when $\pi = 3.142$ and $r = 2.9$

30 If $V = pv^{1.6}$ find V when $v = 6.032$ and $p = 29.12$.

31 If $3^x = 24$, find x.

32 If $R^n = 1.8575$ find R when $n = 18$.

33 When $1 = \dfrac{gt^2}{\pi^2}$ find l when $g = 9.81$, $t = 2$, $\pi = 3.142$.

34 From the formula $V = \sqrt{\dfrac{2ghD}{0.03L}}$ find V when $g = 9.8$, $h = 0.627$, $L = 175$, $D = 0.27$.

35 If $t = 2\pi \sqrt{\dfrac{l}{g}}$, find g when $l = 159$ mm, $t = 0.8$ s, $\pi = 3.142$.

36 Without using tables find the values of:

　　(*a*) $\log 27 \div \log 3$ 　　　　(*b*) $(\log 16 - \log 2) \div \log 2$

37 If $M = PR^n$ find M when $P = 200$, $R = 1.05$, $n = 20$.

38 Find the radius of a sphere whose volume is 500 mm². $V = \frac{4}{3}\pi r^3$.

39 Find the values of

    (1) $\log_e 4.6$ (2) $\log_e 0.062$

40 The insulating resistance, R, of a wire of length l is given by

$$R = \frac{0.42S}{l} \times \log_e \frac{d_2}{d_1} \,.$$

Find l when $S = 2000$, $R = 0.44$, $d_2 = 0.3$, $d_1 = 0.16$.

41 In a calculation on the dryness of steam the following formula was used:

$$\frac{qL}{T} = \frac{q_1 L_1}{T_1} + \log_3 \frac{T_1}{T} \,.$$

Find q when $L_1 = 850$, $L = 1000$, $T_1 = 780$, $T = 650$, $q_1 = 1$.

20

Ratio and Proportion

171 Meaning of a ratio

There are two ways of comparing the magnitudes of two numbers:

(1) By subtraction. This operation states by how much one number is greater or less than the other. If the numbers are represented by a and b, the comparison is expressed by a − b.
(2) By division. By this means we learn what multiple or what part or parts one number is of the other.

The latter is called the ratio of the two numbers, and may be expressed by a ÷ b, or $\frac{a}{b}$, or in the special form a : b. Of these the fractional form is best suited for manipulation.

172 Ratio of two quantities

The magnitude of two quantities of the same kind, such as two lengths, weights, sums of money, etc., may be compared by means of a ratio. To do this the measures of the two quantities are expressed in terms of the same unit by numbers. The ratio of these two numbers expresses the ratio of the quantities.

Thus, the ratio of two distances which are respectively a metres and b metres would be $\frac{a}{b}$ or a : b.

The ratio would be unaltered in value if the quantities were expressed in other units.

Thus the ratio of 3 hours to 2 hours is the same as the ratio of 9 metres to 6 metres or 108° to 72°.

This is obvious from the consideration that, as a ratio can be expressed by a fraction, it can be manipulated in the same ways as fractions.

Thus
$$\frac{a}{b} = \frac{a \times m}{b \times m}$$

and
$$\frac{a}{b} = \frac{a \div m}{b \div m}.$$

A ratio is always a number, either an integer or a fraction (vulgar or decimal), and is not expressed in terms of any particular unit.

173 Proportion

If four numbers a, b, c, d are so related that the ratios $\frac{a}{b}$ and $\frac{c}{d}$ are equal, the numbers are said to be in proportion.

It follows from the definition of a ratio that a and b must represent the ratio of two quantities of the same kind, while c and d must also represent the ratio of two quantities of the same kind, though not necessarily of the same kind as a and b.

Thus a and b might represent the measures of two weights, while c and d might represent the measures of two costs.

Continued proportion

If a series of numbers a, b, c, d . . . is such that

$$\frac{a}{b} = \frac{b}{c} = \frac{c}{d} = \ldots$$

then these numbers are said to be in continued proportion.

Thus in the series of numbers 2, 6, 18, 54 . . . the ratios

$$\tfrac{2}{6} = \tfrac{6}{18} = \tfrac{18}{54} \ldots$$

Mean proportional   If a, b, c are numbers such that

$$\frac{a}{b} = \frac{b}{c},$$

b is called a mean proportional between a and c.

Then $$b^2 = ac$$
and $$b = \sqrt{ac}.$$

In this way we can find a mean proportional between any two numbers.

174   Theorems on ratio and proportion

The following theorems should be noted:

(1) Let $$\frac{a}{b} = \frac{c}{d}.$$

Then $$\frac{a}{b} \times bd = \frac{c}{d} \times bd.$$

$\therefore$ $$ad = bc.$$

(2) Let $$\frac{a}{b} = \frac{c}{d}.$$

Then from the meaning of a ratio

$$\frac{b}{a} = \frac{d}{c} \text{ (alternando).}$$

(3) Let $$\frac{a}{b} = \frac{c}{d}.$$

Then $$ad = bc \text{ (theorem 1).}$$

$\therefore$ $$\frac{ad}{cd} = \frac{bc}{cd}.$$

$\therefore$ $$\frac{a}{c} = \frac{b}{d} \text{ (invertendo).}$$

(4) Let $$\frac{a}{b} = \frac{c}{d}.$$

Then $$\frac{a}{b} + 1 = \frac{c}{d} + 1.$$

$\therefore$ $$\frac{a + b}{b} = \frac{c + d}{d} \text{ (componendo).}$$

(5) Let $$\frac{a}{b} = \frac{c}{d}.$$

Then $$\frac{a}{b} - 1 = \frac{c}{d} - 1.$$

$\therefore$ $$\frac{a - b}{b} = \frac{c - d}{d} \text{ (dividendo).}$$

(6) By division of (4) by (5)

$$\frac{a + b}{a - b} = \frac{c + d}{c - d}.$$

This is called componendo and dividendo.

175   An illustration from geometry

The following illustration from geometry is given as being of great importance. For a proof of it and for a geometrical treatment of ratio and proportion you are referred to any text-book on geometry.

Similar triangles. Triangles which have their corresponding angles equal are called similar. In such triangles,

the ratios of corresponding sides are equal.

In Fig. 42 the triangles ABC, A'B'C' are similar – i.e.,

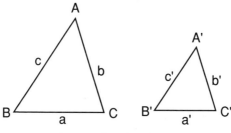

Fig. 42.

$\angle A = \angle A', \angle B = \angle B', \angle C = \angle C'$. Corresponding sides are those which are opposite to equal angles.

Denoting the lengths of the sides opposite to these by a, b, c, a', b', c' as shown, then by the theorem stated above

$$\frac{a}{a'} = \frac{b}{b'} = \frac{c}{c'}.$$

It follows from theorem (3) of Section 174

$$\frac{a}{b} = \frac{a'}{b'}, \frac{b}{c} = \frac{b'}{c'} \text{ and } \frac{a}{c} = \frac{a'}{c'},$$

i.e., the ratios of the pairs of sides containing the equal angles are equal. Each pair of equal ratios gives a set of numbers in proportion.

176 Constant ratios

Let AOB be any angle (Fig. 43).

On the arm OA take points P, Q, R . . . and draw PK, QL, RM perpendicular to the other arm.

Then the △s OPK, OQL, ORM are equiangular and similar.

∴ by the theorem of Section 175

$$\frac{PK}{OK} = \frac{QL}{OL} = \frac{RM}{OM}$$

Clearly any number of points, such as P, Q, R, can be taken and the ratio of all pairs of sides such as the above are equal. We can therefore say that for the angle AOB all such ratios are constant in value.

This constant ratio is called in trigonometry the tangent of the angle AOB. It is abbreviated to tan AOB – i.e.,

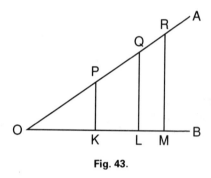

Fig. 43.

$$\frac{PK}{OK} = \frac{QL}{OL} = \frac{RM}{OM} = \tan AOB.$$

A similar result is true for any other angle.

∴ every angle has its own tangent or constant ratio, by which it can be identified.

Referring to Sections 82 and 83 it will be seen that the gradient of a straight line, represented by m, is the tangent of the angle made by the line with the x axis.

In Fig. 24, for example, the ratio $\frac{PQ}{OQ}$ is constant for every point on the line, and is the tangent of the angle POQ.

In the general equation y = mx + b, m represents the tangent of the angle made with the axis of x.

177 Examples of equal ratios

Examples of equal ratios frequently occur in mathematics, and the following theorem, in different forms, is sometimes useful.

Let $\frac{a}{b} = \frac{c}{d} = \frac{e}{f}$ be equal ratios (there may be any number of them).

Let k represent their common value.

Then
$$\frac{a}{b} = \frac{c}{d} = \frac{e}{f} = k.$$

∴
$$a = bk$$
$$c = dk$$
$$e = fk.$$

These results make possible various manipulations.
For example, by addition
$$a + c + e = bk + dk + fk$$
$$= k(b + d + f).$$

∴
$$\frac{a + c + e}{b + d + f} = k,$$

and is therefore equal to each of the original ratio. This can be varied in many ways.

For example, by multiplication of the three equations above:
$$2a = 2bk$$
$$3c = 3dk$$
$$7e = 7fk.$$

∴ by addition
$$2a + 3c - 7e = 2bk + 3ak - 7fk$$
$$= k(2b + 3d - 7f).$$

∴
$$\frac{2a + 3c - 7e}{2b + 3d - 7f} = k = \text{each of the original ratio.}$$

Exercise 64

1 Write down the following ratios:

    (*a*) £a to b pence.   (*b*) p h to q min r s.

2 Write down the ratio of 'a km an hour' to 'a m per second'.

3 If the ratio a:b is equal to the ratio 5:8, find the numerical values of the following ratios:

    (1) $\frac{1}{a}:\frac{1}{b}$ (2) $a^2:b^2$ (3) 2a:3b

4 (*a*) Two numbers are in the ratio of 4:5. If the first is 28, what is the second?

 (*b*) Two numbers are in the ratio of a:b. If the first is x, what is the second?

 (*c*) Two numbers are in the ratio of 1:x. If the second is a, what is the first?

5 If the ratio $\dfrac{a}{b+c}$ is equal to the ratio $\dfrac{c}{x}$, find x.

6 A rectangle of area A mm^2 is divided into two parts in the ratio p:q. Find expressions for their areas.

7 A piece of metal of mass a kg is divided into two parts in the ratio x:y. What are the masses of the parts?

8 If $\dfrac{1}{a}, \dfrac{1}{b}, \dfrac{1}{c}, \dfrac{1}{x}$ are four numbers in proportion, find x.

9 What number must be added to each term of the ratio 11:15, so that it becomes the ratio 7:8?

10 Find the mean proportional between:

 (1) ab and bc   (2) $8a^2$ and $2b^2$
 (3) a(a + b) and b(a + b)

11 What number added to each of the numbers 2, 5, 8, 13 will give four numbers in proportion?

12 If $\dfrac{a}{b} = \frac{5}{2}$, find the value of $\dfrac{a+b}{a-b}$.

13 Find the ratio of $\dfrac{a}{b}$ when

 (1) 3a = 7b
 (2) $16a^2 = 25b^2$ (2) $\dfrac{3}{a} = \dfrac{7}{b}$

14 (1) What is the result when 270 is altered in the ratio 5:3?

 (2) What is the result of altering $\dfrac{a}{b}$ in the ratio 6:5?

 (3) What is the result of altering $\frac{4}{3}$ in the ratio a:b?

15 If a, x, y, b are in continued proportion, find x and y in terms of a and b.

21

Variation

178 Direct variation

In Chapter 16 examples are given of a variable quantity the value of which depended on the value of another variable, and is called a function of it. There are very many different forms which this variation may take, and in this chapter we shall examine one of the most important of them. We will begin with a very simple example.

If a man is paid at a certain rate for every hour that he works, the total amount he earns over any period depends on the number of hours he works. If he doubles the number of hours he works, his earnings will be doubled. Generally the ratio of the amounts he earns in any two periods is equal to the ratio of the number of hours worked in the periods.

If T_1 and T_2 represent the number of hours worked in two periods and if W_1 and W_2 represent the amounts of wages earned in them,

then
$$\frac{W_1}{W_2} = \frac{T_1}{T_2}.$$

These four quantities are in proportion (see Section 173).

Hence, when the relations between the two quantities can be expressed in this way we say that

The wages are proportional to the time worked or the wages vary directly as the time.

The wages are a function of the time, and the words 'proportional to' or 'vary directly as' are used to define the exact functional relations which exist between the two quantities.

Direct variation may be defined as follows:

If two quantities Y and X are so related that $\dfrac{Y_1}{X_1} = \dfrac{Y_2}{X_2}$ *, where X_1 and X_2 are any two values of X and Y_1 and Y_2 are corresponding values of Y, then Y is said to be proportional to X, or Y varies directly as X.*

In order to discover whether or not one quantity varies directly as another, this simple test can be applied.

If one quantity is doubled, is the other doubled in consequence? Or, more precisely, if one quantity is altered in a certain ratio, is the other altered in the same ratio?

The sign $\propto$ is used to denote direct variation. Thus in the case above we would write $Y \propto X$.

179 Examples of direct variation

(1) The distance travelled by a car moving with uniform speed varies directly as the time.
(2) The mass of an amount of water is proportional to the volume.
(3) The circumference of a circle varies directly as the diameter.
(4) The electrical resistance of a wire varies directly as the length.

180 The constant of variation

If $y \propto x$ then $y = kx$, where k is a constant.

In Section 177 the common value of a number of equal ratios was represented by k, a constant number for these ratios.

When one quantity varies directly as another, then we have seen that the ratio of corresponding pairs of values of the variables is constant. Consequently, as in Section 177, this constant is usually represented by k, and by means of it the relation between the quantities can be expressed as a formula.

For instance, in Example (1) of Section 179 it was stated that for a body moving with uniform speed, distance moved varies directly as time.

Let s represent any distance travelled.

Let t represent the corresponding time.

Then, since s varies directly as t, any ratio of corresponding values of these is constant.

Let k represent this constant.

Then by definition $\dfrac{s}{t} = k$.

∴ $s = kt$.

This is the form of the law connecting s and t, but it cannot be of much use until the value of k is known for this particular case.

To find this, we must know a pair of corresponding values of s and t. Thus if we are told that the car moves 40 m in 2.5 s, then, substituting the values fo s and t, we get

$$40 = k \times 2.5.$$
∴ $$k = 40 \div 2.5$$
$$= 16.$$

∴ the law connecting s and t for this particular velocity is

$$s = 16t.$$

181   Graphical representation

If x and y represent two variables such that $y \propto x$, then, as shown in Section 180, $y = kx$.

The form of this equation is the same as $y = mx$, the graph of which was shown in Section 82 to be a straight line passing through the origin, m being the gradient of the line.

∴ $y = kx$ represents a straight line of gradient equal to k.

Consequently the graphical representation of the variation of two quantities where one varies directly as the other is a straight line passing through the origin.

182   To find the law connecting two variables

The engineer and the scientist frequently require to know the law connecting two variables, corresponding values of which have been found by experiments. The law may assume many forms, but in some cases there may be reason to suppose that one of the quantities varies directly as the other – i.e., the law may be of the form $y = kx$.

From the results of the experiment we can proceed to determine:

(1) Is the law one of direct variation?

(2) If it is, and the law is of the form y = kx, the value of k must be found.

To determine (1) the results are plotted. Then:

(*a*) If the graph is a straight line, the law is one of direct variation.
(*b*) If the straight line passes through the origin, the equation is of the form y = kx.

Then k has to be determined.

Graphically, k can be found as in Section 176 by finding the tangent of the angle made with the axis of x.

Algebraically as shown in Section 180. A pair of corresponding values of x and y is chosen from the graph. These are substituted in y = kx, and so k is determined.

183   Worked example

A spiral spring was extended by hanging various weights from it.

The amounts of extension of the spring for different weights were observed and tabulated as follows:

Weight in N	0	0.1	0.2	0.3	0.4	0.5
Extension in cm	0	0.15	0.3	0.44	0.6	0.75

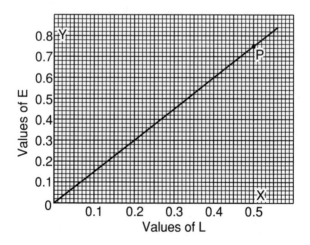

Fig. 44.

From these results discover the law connecting the attached weight and the extension of the spring.

The graph resulting from plotting these results is shown in Fig. 44. It is a straight line passing through the origin though one of the points, corresponding to a weight of 0.3 N, lies slightly off the line. This is to be expected in experimental results. Also the line passes through the origin.

∴ as shown in Section 182, the law connecting the weight and extension is one of direct variation – i.e.,

The extension varies directly as the attached load.

Let E = extension
Let L = load

then E ∝ L and E = kL
To find k a pair of values is taken, at the point P where
E ∝ 0.75 and L = 0.5.

Substituting these   0.75 = k × 0.5
whence   k = 1.5.
∴ the law is   E = 1.5L.

184   y partly constant and partly varying as x

The case frequently occurs, in practical work, of a variable quantity which in part varies directly as another quantity and in part is constant.

There is an example of this in Section 77 in the problem concerning the profit of a restaurant. The profit depends on

(1) The number of customers which is variable.
(2) The overhead charges which are constant.

It was found that y = ax − b was the law which connected these, where y, the profit, varies directly as x, the number of customers, and also depends on the constant charges b.

In general all such cases can be represented by the equation y = kx + b.

This is the equation of a straight line, but it does not pass through the origin (Section 83). It contains the two constants k and b, which must be determined before the law connecting x and y can be stated.

If two pairs of values of x and y are known, the solution can be found as shown in Section 65, example 2.

In practical work pairs of values are found by experiment. The worked example which follows shows the method of procedure in such cases.

185   Worked example

When two voltmeters are compared they have readings corresponding to C and K below.

C	1.9	2.75	3.8	4.8	5.8
K	5.75	8.3	11.2	14	16.8

It is thought that C and K are connected by a law of the form

$$K = mC + b.$$

Test this by plotting the points and find the values of m and b.

The law $K = mC + b$ is linear – i.e., it is in the general form of the equation of a straight line. To test them we must find if the plotted points be on a straight line.

Comparison of the two sets of values suggests that the scale units for C on the axis of x should be larger than those for K on the y axis.

When the points are plotted they are seen to lie approximately on a straight line, slight deviations being due to experimental

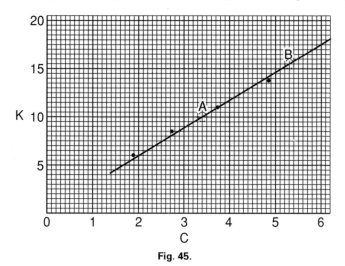

Fig. 45.

errors. When this line is drawn as evenly as possible, it appears as in Fig. 45.

Two suitable points, A and B, are selected on the line, and their co-ordinates are as follows:

For A, C = 3.4, K = 10.
For B, C = 5.3, K = 15.5.

These are to satisfy the equation

$$K = mC + b.$$

Substituting we get:

$$10 = 3.4m + b \qquad (1)$$
$$15.5 = 5.3m + b \qquad (2)$$

Subtracting, $1.9m = 5.5$

∴ $m = \dfrac{5.5}{1.9} = 2.9$ approx.

Substituting for m in (1),

$$10 = (3.4 \times 2.9) + b.$$
∴ $b = 10 - 9.86$
$$= 0.14.$$

∴ the law is $K = 2.9C + 0.14.$

Exercise 65

1 The following are examples in which the value of one quantity depends on another. State in each case whether or not it is a case of direct variation:

   (*a*) Distance and time when a man runs the 1000 m race.
   (*b*) Interest and time when money bears interest at a fixed rate.
   (*c*) The logarithm of a number and the number itself.
   (*d*) The y co-ordinate of a point on a straight line and the x co-ordinate.
   (*e*) The cost of running a school and the number of scholars.

2 If y = kx and y = 8 when x = 7, find k. Hence find y when x = 40.

3 If y is proportional to x and y = 10 when x = 4, find y when x = 15; also find x when y = 8.4.

4 If y $\propto$ x, and when y = 16.5, x = 3.5, find the law connecting x and y. Hence find x when y = 21.

5 The distances travelled by a body from rest were as follows:

Time (s)	1	2	3	4
Distance (m)	3.2	6.4	9.6	12.8

Plot these and find if distance varies directly as time. If it does, find:

(1) The law connecting time (t) and distance (s).
(2) The distance passed over in 2.8 s.

6 The extension (E) of a spring varies directly as the force (W) by which it is stretched. A certain spring extended 2.4 when stretched by a weight of 4.5. Find:

(1) The law which connects them.
(2) The extension due to a weight of 7.

7 The law connecting two variables x and y is of the form y = kx + b, where k and b are constants. When x = 10, y = 11, and when x = 18, y = 15. Find k and b and state the law.

8 In a certain machine the law connecting the applied force (E) and the load (W) was of the form E = aW + b, where a and b are constants. When w = 20, E = 1.4, and when W = 30, E = 2. Find the law.

9 In experiments to determine the friction (F N) between two metallic surfaces when the load is W N, the following results were found:

W	3	5	7	10	12
F	0.62	1.5	2.4	3.6	4.4

Assuming W and F to be connected by a law of the form F = aW + b, find the law by drawing the average straight line between the points.

186 y varies as the square of x – i.e., y $\propto$ x^2

If the sides of a square be doubled, the area is not doubled, but is four times as great; if the sides be trebled the area is nine times as great. The area of a square varies directly as the square of the length of its sides.

The ratio of the area of a circle to the square of the length of its radius can be shown experimentally to be constant. If A = the area and r = radius, then $\dfrac{A}{r^2}$ is constant for all circles. This constant is represented by the special symbol π, and its value is approximately 3.1416.

$\therefore$ the area of a circle varies directly as the square of the length of its radius.

Students of mechanics will know that the distance passed over by a body moving with uniformly increasing velocity is proportional to the square of the time. A special case is a falling body.

If s = the distance fallen and t = the time taken in s, then s $\propto$ t^2 and s = kt^2.

Experiments show that k = $\frac{1}{2}$g, where g is an absolute constant whose value is approximately 0.81 m/s^2.

The graphical representation of y = kx^2 is that of a quadratic function (see Chapter 16). For different values of k, the graph is a parabola, symmetrical about the y axis and having its lowest point at the origin (Section 125).

187 y varies as the cube of x – i.e., y $\propto$ x^3

If the edge of a cube be doubled, the volume is 8 times as great. The volume varies directly as the cube of the length of an edge.

The volume of a sphere also varies as the cube of the radius.

If V = the volume and r = the length of the radius, then V $\propto$ r^3 and V = kr^3.

It can be shown that k = $\frac{4}{3}\pi$.

$\therefore$ $$V = \tfrac{4}{3}\pi r^3.$$

The graph of y = x^3 is a curve, as shown in Fig. 46. It is called a cubical parabola.

You should make a table of values and draw the curve.

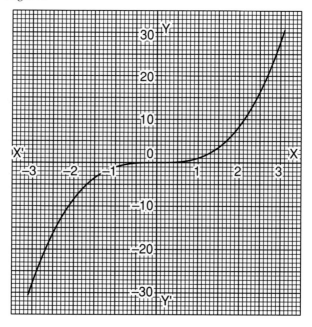

Fig. 46.

188 y varies as $\sqrt{x}$ or $x^{\frac{1}{2}}$, i.e., $y \propto \sqrt{x}$

This form of variation, besides arising in various physical examples, may also be regarded as the inverse of $y = x^2$.

Since
$$y = kx^2$$
$$x^2 = \frac{y}{k}.$$

∴
$$x = \sqrt{\frac{1}{k}} \times \sqrt{y}.$$

Since $\sqrt{\frac{1}{k}}$ is a constant, then $x \propto \sqrt{y}$.

The graph of $y = \sqrt{x}$ is a parabola, as in Fig. 47. It is the same curve as Fig. 31, as will be realised from Sections 122 and 123, but being the inverse, it is symmetrical about the x axis. Negative roots will be found on the part of the curve below the x axis.

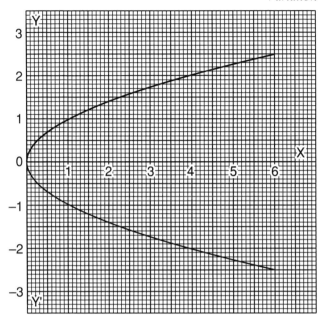

Fig. 47.

189 Inverse variation: $y \propto \dfrac{1}{x}$

Let x and y be two numbers such that their product is constant – i.e., xy = k,

then $\qquad\qquad y = \dfrac{k}{x}$ or $x = \dfrac{k}{y}$.

Each quantity can be expressed in terms of the reciprocal or inverse of the other.

Many examples of this occur in mathematics. The following is a simple case.

Let x and y be the sides of a rectangle of area 60 m^2, then xy = 60.

The lengths of the sides may be varied in very many ways, their product always being equal to 60. If x is increased, y will be decreased, and vice versa. If x be doubled, y will be halved. In general, if x be changed in a given ratio, y will be changed in the inverse ratio.

Hence we say that y varies inversely as x – i.e., $y \propto \dfrac{1}{x}$.

Hence $\qquad\qquad\qquad y = \dfrac{k}{x}$.

Among the many examples of inverse variation we may note:

(1) Time to travel a given distance varies inversely as the speed. If the speed be doubled, the time is halved.
(2) The volume of a fixed mass of gas varies inversely as the pressure on it, the temperature remaining constant.
    If p = the pressure and v = the volume.

$$p \propto \frac{1}{v} \text{ or } p = \frac{k}{v} .$$

(3) The electrical resistance of a wire of given length and material to the passage of a current through it varies inversely as the area of the cross-section of the wire.
    If R = the resistance,
    and A = the area of the cross section,

then $\qquad\qquad\qquad R \propto \dfrac{1}{A} \text{ or } R = \dfrac{k}{A}$.

190   Graph of $y = \dfrac{k}{x}$

In its simplest form when k = 1, the equation becomes $y = \dfrac{1}{x}$.

The graph of this function presents some new difficulties which will be apparent on drawing the curve. The following is a table of values formed in the usual way:

x	$\frac{1}{4}$	$\frac{1}{2}$	1	2	3	4
y	4	2	1	$\frac{1}{2}$	$\frac{1}{3}$	$\frac{1}{4}$

A similar set of values can be tabulated for negative values of x, the corresponding values of y being negative.

The curve, which is called a hyperbola, is shown in Fig. 48. It consists of two branches, alike in shape, one for + ve values and

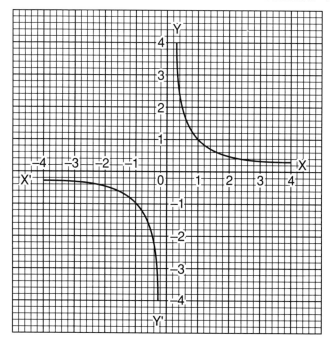

Fig. 48.

the other for $-$ve values of x. The following important features of this curve should be noted.

(1) As x increases, y or $\frac{1}{x}$ decreases. When x becomes very great, y becomes very small, and the curve approaches very close to the axis of x.

(2) As x decreases, y or $\frac{1}{x}$ increases. When x is very small, y is very large, and the curve approaches very close to the axis of y. Both these features of the curve are repeated for $-$ve values of x. It may be noted that the curve is symmetrical (i) about the line through the origin making $45°$ with the x axis – i.e., the line $y = x$; (ii) about a line through the origin at right angles to this – i.e., the line $y = -x$ (see Sections 82 and 83).

For other values of k the curve of $y = \frac{k}{x}$ is always a hyperbola.

191 Other forms of inverse variation

(1) One quantity may vary inversely as the square of another quantity – i.e., $y \propto \dfrac{1}{x^2}$, whence $y = \dfrac{k}{x^2}$. In electricity, for example, the force between two magnetic poles varies inversely as the square of the distance between them. Many other physical laws involve this form of variation.

(2) Another form of variation is that in which one quantity varies inversely as the cube of another – i.e., $y \propto \dfrac{1}{x^3}$ and $y = \dfrac{k}{x^3}$.

Generalising, y may vary directly or inversely as any power of x – i.e., $y \propto x^n$ or $y \propto \dfrac{1}{x^n}$.

In all cases of direct variation the same method, the introduction of the constant k, is followed and the evaluation of k proceeds on the same lines.

192 Worked examples

Example 1: If y varies as the cube root of x, and if $y = 3$ when $x = 64$, find the formula connecting the variables. Hence find x when $y = \frac{15}{4}$.

(1) Since
$$y \propto x^{\frac{1}{3}}$$
$$y = kx^{\frac{1}{3}}.$$
$$\therefore \quad k = y \div x^{\frac{1}{3}}.$$

Substituting the given values
$$y = 3 \div 64^{\frac{1}{3}}$$
$$= 3 \div 4 \text{ or } \tfrac{3}{4}.$$
$\therefore$ the law is $\quad y = \tfrac{3}{4}\sqrt[3]{x}.$

(2) When $y = \frac{15}{4}$, we have
$$\tfrac{15}{4} = \tfrac{3}{4}\sqrt[3]{x}.$$
$$\therefore \quad \sqrt[3]{x} = \tfrac{15}{4} \div \tfrac{3}{4}.$$
$$= 5.$$
$$\therefore \quad x = 5^3 = 125.$$

Example 2: The time of vibration of a simple pendulum varies as the square root of its length. If the length of a pendulum which beats seconds is 1 m, what will be the time of vibration if its length is increased by 6 cm?

Let l = length of the pendulum in metres and
let t = the time of variation.

Then $\qquad t \propto \sqrt{l}$ and $t = k\sqrt{l}$.

When l = 1.00, t = 1.00 and $1 = k\sqrt{1}$, $\therefore$ k = 1.

When l = 1.06, $t = k\sqrt{l} = 1\sqrt{1.06} = 1.03$.

$\therefore \qquad\qquad\qquad t = 1.03$ s.

Exercise 66

1 If y is proportional to x^2 and when x = 15, y = 200, find the equation connecting x and y. Find y when x = 8.5.

2 If $y \propto x^3$ and x = 2 when y = 2, find the law connecting x and y. Then find y when x = 3.

3 If $y \propto \dfrac{1}{x^2}$, fill up the blanks in the following table:

x	1	2	3	
y			$\frac{4}{3}$	$\frac{1}{3}$

4 If $y \propto \sqrt{x}$ and if y = 3.5 when x = 4, express y in terms of x. What is y when x = 25?

5 If $y \propto x^3$ and if y = 6 when x = 4, find the value of y when x = 16. Also find x when y = 3.

6 If y is inversely proportional to x, and x = 5 when y = 6, find the law connecting x and y and find x when y = 20.

7 If $y \propto \dfrac{1}{x}$, fill up the gaps in the following table:

x		1.2	8	
y	6		1.5	0.8

8 The force F which acts between two magnetic poles is inversely proportional to the square of the distance (d) between them. Express this as a formula if F = 120 when d = 4.

9   The distance through which a heavy body falls from rest varies as the square of the time taken. A lead ball falls through 490 m in 10 s. How long will it take to fall through 78.4 m?

10   The intensity of illumination given by a projector varies inversely as the square of the distance of its lamp from the screen. If the projector is 20 m from the screen, where must it be placed so that the illumination is 4 times as great?

11   If the distances of an object and of its image, formed by a mirror, are measured from a certain point, it is found that the sum of the distance varies as their product. If the image distance is 120 mm when the object distance is 300 mm, calculate the image distance when the object distance is 540 mm.

12   The square of the speed of a particle varies as the cube of its distance from a fixed point. If this distance is increased by 1.2 per cent, what is the approximate percentage increase in the speed?

13   A clock keeps accurate time at 10°C, but gains as the temperature falls and vice-versa, the rate of gain or loss varying as the square of the number of °C between the actual temperature and 10°C. If it gains 2 s per day when the temperature is 5°C, how much does it lose (to the nearest s) in 4 days when the temperature is 42°C?

14   If $y \propto x^{1.4}$ and if $y = 354.5$ when $x = 15$, find the law connecting y and x.

193   Functions of more than one variable

Example 1: It is proved in elementary geometry that the area of a triangle is given by the formula

$$A = \tfrac{1}{2}bh$$

where A = the area of the triangle
      b = the length of a base
      h = the corresponding altitude.

Both b and h are variables, and the value of A depends on them both.

∴ A is a function of the two variables b and h.

In any triangle let the height remain constant, but the base variable, then if the base be doubled, the area will be doubled. If the base remain constant and the height be trebled, the area will be trebled.

Now suppose both base and height to vary; let the base be doubled and the height trebled, then the area will be 2×3 – i.e., 6 times greater.

We can infer, then, that if both base and height vary, the area varies as the product of base and height – i.e.,

$$A \propto b \times h.$$
$$\therefore \qquad A = kbh.$$

From geometrical considerations we know that in this case $k = \frac{1}{2}$.

Example 2: In Section 189 it was stated that the volume of a given mass of gas, at a constant temperature, varies inversely as the pressure on it. It can also be shown by experiment that if the pressure be kept constant and the temperature varied, then the volume varies directly as the absolute temperature. If both temperature and pressure vary, then the volume varies directly as the absolute temperature and inversely as the pressure.

Let v = volume
$\quad\;\; T$ = absolute temperature
$\quad\;\; p$ = pressure

then
$$v \propto \frac{T}{p}.$$

$$\therefore \qquad v = k \times \frac{T}{p}.$$

Example 3: If an electric current passes throgh a wire, it encounters resistance. This resistance varies as the length of the wire (Section 179), and in a wire of given length – i.e., constant – it varies inversely as the cross section of the wire (Section 189).

$\therefore$ if R = the resistance
$\quad\;\; l$ = the length
$\quad\;\; A$ = the cross section

then R varies directly as l and inversely as A.

Thus
$$R \propto \frac{l}{A}$$

and
$$R = k \times \frac{l}{A}.$$

194   Joint variation

The variation of a quantity due to two or more variables is sometimes called joint variation, and the quantity is said to vary jointly as their product.

In dealing with problems involving joint variation, the same procedure with regard to the constant of variation and its determination is followed as in the previous case. The following examples will serve to illustrate it.

195   Worked example

Example 1: A quantity represented by y varies directly as x and inversely as z^3. It is known that when x = 15, z = 12 and y = $\frac{1}{36}$. Find the law connecting the quantities.

We are given that $y \propto \dfrac{x}{z^3}$.

$\therefore$ $y = k \times \dfrac{x}{z^3}$.

Substituting the given values

$$\tfrac{1}{36} = k \times \frac{15}{12^3} .$$

$\therefore$ $k = \tfrac{1}{36} \div \dfrac{15}{12^3} = \tfrac{16}{5}$.

$\therefore$ the law is $y = \tfrac{16}{5} \dfrac{x}{z^3}$.

Example 2: The force between two magnetic poles varies jointly as the product of their strength and inversely as the square of the distance between them. If two poles of strength 8 and 6 units repel one another with a force of 3 N when placed 4 m apart, with what force will two poles whose strengths are 5 and 9 units repel one another when 2 m apart?

Let   F = the force
m_1, m_2 = the pole strengths
d = the distance apart

then $F \propto \dfrac{m_1 m_2}{d^2}$.

$\therefore$ $F = k \times \dfrac{m_1 m_2}{d^2}$.

Substituting the given values

$$3 = k \times \frac{8 \times 6}{4^2}.$$

$$\therefore \qquad k = \frac{3 \times 4^2}{8 \times 6} = 1.$$

$$\therefore \qquad F = \frac{m_1 m_2}{d^2}.$$

In the second case $F = \frac{5 \times 9}{4} = 11.25$ N.

Exercise 67

1 Express the following statements in the form of equations:

    (a) y varies jointly as x and z.
    (b) y varies directly as x and inversely as the square of z.
    (c) y varies directly as the square root of x and inversely as z.
    (d) The volume (V) of a cylinder varies jointly as the height (h) and the square of the radius of the base (r).
    (e) The weight (W) which can be carried safely by a beam varies inversely as the length (l), directly as the breadth (b) and directly as the square of the depth (d).
    (f) y varies directly as the square of x and inversely as the cube root of z.

2 If y varies directly as x and inversely as z, and if $y = 10$ when $x = 8$ and $z = 5$, find the law connecting x, y and z. Also find y when $x = 6$ and $z = 2.5$.

3 If y varies jointly as x and z^2, and if $y = 13\frac{1}{3}$ when $x = 2.5$ and $z = \frac{4}{3}$, find the law connecting the variables. Also find x when $z = \frac{3}{2}$ and $y = 54$.

4 y varies directly as x^2 and inversely as $\sqrt{z}$. When $x = 8$ and $z = 25$, $y = 16$. Find y when $x = 5$ and $z = 9$.

5 The load that a beam of given depth can carry is directly proportional to the breadth and inversely proportional to the length. If a beam of length 7 m and width 175 mm can support a load of 4 t, what load can be supported by a beam of the same material 5 m long and 250 mm wide?

6 If z varies as x^2 and inversely as y^2, and if $z = 4$ when $x = 8$ and $y = -0.5$, find z when $x = -2$ and $y = 0.25$.

7 The number of heat units (H) generated by an electric current varies directly as the time (t) and the square of the voltage (E)

and inversely as the resistance (R). If H = 60, when t = 1, E = 100 and R = 40 find the law connecting them.
Also find:

(1) the value of H when E = 200, R = 120 and t = 300.
(2) The value of t when E = 120, R = 90 and H = 5760.

8   The pressure in a test apparatus varies directly as the absolute temperature and inversely as the volume of gas. If the temperature increases by 2 per cent and the volume decreases by 2 per cent, what is the change in the pressure?

22

The Determination of Laws

196 Laws which are not linear

In the preceding chapter we considered the determinations of laws which were linear, and which were arrived at by using experimental data.

But such laws are not always represented by straight lines. They may involve powers of a variable such as were considered in Section 186 and onwards. In these cases, when the results of the experiments are plotted, they will lie on a portion of a curve which might be one of those illustrated in the preceding chapter or of many others. In practice, when only a small portion of the curve can be drawn, it is impossible to identify what curve it is.

There are two devices, however, by means of which a straight line can be obtained instead of a curve. The identification can then be made by the methods previously considered.

197 $y = ax^n + b$. Plotting against a power of a number

If the law which we require to find is of the form $y = ax^n + b$ or $y = ax^n$, where n is known, then we plot corresponding values of y and x^n instead of y and x. The resulting graph will be a straight line.

Let us consider as an example $y = 2x^2$.

The graph of this function was found in Section 125 to be a parabola. The values given to x were 0, 1, 2, 3 . . .

If we plot corresponding values of y and x^2 the table of values will be

x^2	0	1	2	3	4
y	0	2	4	6	8

and so on. The values of x are not shown in the table.

The resulting graph is a straight line passing through the origin as shown in Fig. 49. It is the same as $y = 2x$ when y is plotted against x.

If the equation is $y = 2x^2 + 3$ on plotting in the same way, that is y against x^2, the resulting line will not pass through the origin but will have an intercept of 3 units on the y axis. It will be the same as $y = 2x + 3$, when y is plotted against x.

The same procedure will be followed with any other power of x. Thus in $y = 2x^3 + 5$, y is plotted against x^3.

In general for the function $y = ax^n + b$ plot y against x^n.

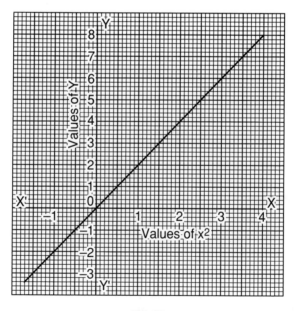

Fig. 49.

If $x^n = z$ then the equation takes the form of

$$y = az + b.$$

This is the equation of a straight line. The graph, while not showing the relation between y and x as graphs usually do, will make it possible to find the values of a and b by methods previously given.

Worked example

Two variables, x and y, are thought to be connected by a law of the form of $y = ax^2 + b$. The following values of x and y are known. Find the law connecting the variable.

x	0.5	1	1.5	2	2.5
y	-9.25	-7	-3.25	2	8.75

y must be plotted against x^2; we therefore calculate the following tables of corresponding values of x^2 and y.

x^2	0.25	1	2.25	4	6.25
y	-9.25	-7	-3.25	2	8.75

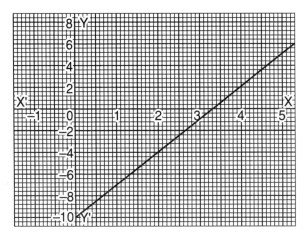

Fig. 50.

Plotting y against x^2, the resulting graph is as shown in Fig. 50.

This is a straight line, and the values of a and b can be found by the method of Section 185.

By inspection of the graph, the intercept on the y axis (i.e., b) is -10, and a, the gradient of the line, is 3.

$\therefore$ the law is $\qquad y = 3x^2 - 10.$

198 $y = ax^n$. Use of logarithms

As was pointed out, the previous method can be used only when the power of x involved is known. If, however, it is not known, and the law is of the form $y = ax^n$, this can be reduced to the form of the equation of a straight line by taking logarithms.

If $\qquad\qquad y = ax^n$
then $\qquad\qquad \log y = n \log x + \log a.$

Comparing this with the standard form of the equation of a straight line, viz., $y = ax + b$, it is seen to be linear and of the same form, log y taking the place of y, and log x taking the place of x.

The constants to be determined are now n and log a. Therefore we plot the graph of

$$\log y = n \log x + \log a.$$

From this graph n and log a can be found in the same way as a and b in the standard form. When log a is known we find a from the tables and the law can be written down.

It is possible to deal with only the simpler cases in this book.

199 Worked example

Two variables, x and y, are connected by a law of the form $y = ax^n$. The following table gives corresponding values of x and y. Find the law connecting these.

x	18	20	22	24	25
y	623	863	1160	1519	1724

Since the law connecting these is of the form

$$y = ax^n$$

taking logs $\qquad \log y = n \log x + \log a.$

Tabulating values of log x and log y, we get the following:

log x	1.255	1.301	1.342	1.380	1.398
log y	2.795	2.936	3.065	3.182	3.236

The logs correspond in order to the numbers in the column above, and are calculated approximately to 3 places of decimals. The graph is the straight line shown in Fig. 51.

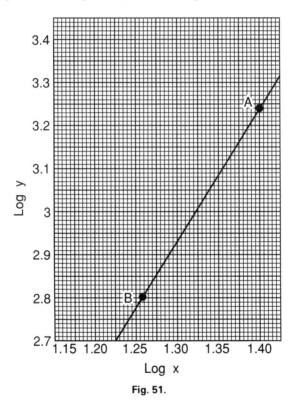

Fig. 51.

Selecting the points A and B on the straight line, we substitute their co-ordinates in turn in the equation

$$\log y = n \log x + \log a.$$

Thus we get the equations

$$3.236 = 1.398n + \log a$$
$$2.795 = 1.255n + \log a.$$

Subtracting, $\quad 0.441 = 0.143n.$

$$\therefore \qquad n = \frac{0.441}{0.143}$$
$$= 3.1 \text{ approx.}$$

Substituting for n in equation above

$$2.795 = (3.1 \times 1.255) + \log a.$$
$$\therefore \qquad \log a = 1.795 - 3.891$$
$$= \overline{2}.904.$$

$\therefore$ a = 0.08 approx.

$\therefore$ the law connecting y and x is

$$y = 0.08x^{3.1}.$$

Exercise 68

1 The variables x and y are connected by a law of the form $y = ax^2 + b$. The following corresponding values of x and y are known. Find the law.

x	0.5	1	1.5	2	2.5
y	4.5	9	16.5	27	40.5

2 The following table gives related values of x and y. Determine whether these values are connected by an equation of the form $y = ax^2 + b$ and, if so, find the values of a and b.

x	4	5	6	7	8	9
y	14.3	18	22.5	28	34.5	41.5

3 The following values of R and V are possible connected by a law of the type $R = aV^2 + b$. Test if this is so and find the law.

V	12	16	20	22	24	26
R	6.44	7.56	9	9.84	10.76	11.76

4 H is connected with V by an equation of the form $H = aV^3 + b$. The following corresponding values are known. Find the values of a and b.

V	10	12	14	15	17
H	1500	2300	3400	4000	5700

5 The following corresponding values of x and y were measured. There may be errors of observation. Test if there is a probable law $y = a + bx^2$ and, if this is the case, find the probable values of a and b.

x	1	1.5	2	2.30	2.50	2.70	2.80
y	0.77	1.05	1.50	1.77	2.03	2.25	2.42

6 In measuring the resistance, R ohms, of a carbon-filament lamp at various voltages, V, the following results were obtained:

V	60	70	80	90	100	120
R	70	67.2	65	63.3	62	60

The law is thought to be of the form $R = \dfrac{a}{V} + b$. Test this and find a and b.

7 The values of x and y in the following table are connected by a law of the form $y = ax^n$. Find a and n and so determine the law.

x	2	3	4	5
y	2	6.75	16	31.25

8 The following table gives corresponding values of two variables x and y. The law which connects them is of the form $y = ax^n$. Find this law.

x	2	3	5	6	8	10
y	4.24	5.20	6.71	7.35	8.49	9.49

9 The following values of H and Q are connected by a law of the type $Q = aH^n$. Find a and n.

H	1.2	1.6	2.0	2.2	2.5	3
Q	6.087	6.751	7.316	7.571	7.927	8.467

10 Two quantities x and y are connected by an equation of the form $y = ax^n$. The following table gives corresponding values of the variables. Determine a and n.

x	3	3.5	4	4.5	5
y	6.19	6.79	7.35	7.89	8.40

23

Rational and Irrational Numbers.
Surds

200 Rational and irrational numbers

A number which is either an integer or can be expressed as the ratio of two integers is called a rational number.

A number which cannot be expressed as an integer or as a ratio with a finite number of figures is called an irrational number.

Thus $\sqrt{2}$ cannot be expressed as a fraction or a ratio with a finite number of figures. It can, however, be expressed as a decimal to any required degree of accuracy.

Thus to 4 significant figures $\sqrt{2} = 1.414$.

6                $\sqrt{2} = 1.414\ 21$ and so on.

But it can be proved that there is no limit to the number of figures in the decimal place, that is to say the number of figures is not finite. It is therefore an irrational number.

Other roots such as $\sqrt{3}$, $\sqrt{11}$, $\sqrt{19}$, $\sqrt[3]{4}$, . . . are examples of irrational numbers. A root of a rational number which is irrational as these are is called a surd.

There are also other numbers, which do not involve roots which are irrational. Thus the ratio of the circumference of a circle to its diameter, which we denote by the symbol π, cannot be expressed by a ratio with a finite number of figures. It is often expressed roughly by $\frac{22}{7}$, or 3.1416 to 5 significant figures, but it has been proved that there is no limit to the number of figures in the decimal part. Computers have evaluated π to an immense number of places, but they are, of course, no nearer an end.

Such numbers as this are irrational, but not surds. They are also called incommensurable.

201 Irrational numbers and the number scale

Since such numbers cannot be expressed with absolute accuracy it is not possible to assign an exact position for them in the complete number scale (Section 38). We can, however, state to any required degree of accuracy the limits between which they lie. Thus $\sqrt{2}$ lies on the scale between 1.414 and 1.415.

or more accurately between 1.4142 and 1.4143
or more accurately still between 1.414 21 and 1.414 22
and so on.

202 Geometrical representation of surds

It may be noted, however, that it is theoretically possible in many cases to obtain by geometrical constructions, straight lines which do represent surds accurately.

For example, we know from geometry that the square on the hypotenuse of a right-angled triangle is equal to the sum of the squares on the sides containing the right angle. Consequently if a right-angled triangle be constructed, the sides of which are of unit length, the length of the hypotenuse must be $\sqrt{1^2 + 1^2}$ – i.e., $\sqrt{2}$.

Similarly the hypotenuse of a right-angled triangle of sides 1, and $\sqrt{2}$ units will be $\sqrt{3}$ units. In this way it is possible to represent many surds by straight lines. The lengths thus obtained can be marked on the number scale, but in practice no high degree of accuracy can be obtained in the construction of them.

203 Operations with surds

It is the custom in algebra to classify such a number as $\sqrt{a}$ as a surd, though until a numerical value has been assigned to a we cannot say whether or not it is irrational. For purposes of operation, however, it is treated as a surd.

In operating with surds one principle is fundamental. Surds must obey the laws of algebra as formulated for rational numbers. Since surds can also be written as powers with fractional indices, e.g. $\sqrt{2} = 2^{\frac{1}{2}}$, we can operate with surds as with these powers, according to the laws of indices.

For example, just as $(a + b)^{\frac{1}{2}}$ is not equal to $a^{\frac{1}{2}} + b^{\frac{1}{2}}$, so $\sqrt{a + b}$ is not equal to $\sqrt{a} + \sqrt{b}$. In this respect the root sign has the same effect as a bracket; the expression under it must be regarded as a whole.

(1) Multiplication

$$\sqrt{a} \times \sqrt{b} = a^{\frac{1}{2}} \times b^{\frac{1}{2}} = (ab)^{\frac{1}{2}} = \sqrt{a \times b}.$$

Thus:

$$\sqrt{3} \times \sqrt{7} = \sqrt{3 \times 7} = \sqrt{21}$$
$$\sqrt{2}(\sqrt{5} - \sqrt{3}) = (\sqrt{2} \times \sqrt{5}) - (\sqrt{2} \times \sqrt{3}) = \sqrt{10} - \sqrt{6}$$
$$4\sqrt{7} = \sqrt{16} \times \sqrt{7} = \sqrt{16 \times 7} = \sqrt{112}$$
$$3x^2\sqrt{y} = \sqrt{9x^4} \times \sqrt{y} = \sqrt{9x^4 y}$$
$$(\sqrt{a} + \sqrt{b}) = (\sqrt{a})^2 + 2(\sqrt{a} \times \sqrt{b}) + (\sqrt{b})^2$$
$$= a + b + 2\sqrt{ab}$$
$$(\sqrt{a} - \sqrt{b})^2 = a + b - 2\sqrt{ab}$$
$$(\sqrt{a} + \sqrt{b})(\sqrt{a} - \sqrt{b}) = (\sqrt{a})^2 - (\sqrt{b})^2 = a - b$$
$$(\sqrt{5} + 7)(\sqrt{3} - 2) = (\sqrt{5} \times \sqrt{3}) - 2\sqrt{5} + 7\sqrt{3} - 14.$$

By using the above and the converse rules we obtain useful transformations in operations.

Thus:

$$\sqrt{1000} = \sqrt{100 \times 10} = \sqrt{100} \times \sqrt{10} = 10\sqrt{10}$$
$$\sqrt{72} = \sqrt{36 \times 2} = \sqrt{36} \times \sqrt{2} = 6\sqrt{2}$$
$$\sqrt{9a^3b^2} = \sqrt{9 \times a^2 \times a \times b^2} = 3ab\sqrt{a}.$$

The above transformations may also be employed to simplify expressions involving surds.

Examples

(1) $\sqrt{5} + \sqrt{20} = \sqrt{5} + \sqrt{4 \times 5} = \sqrt{5} + 2\sqrt{5} = 3\sqrt{5}.$
(2) $\sqrt{27} - \sqrt{75} + \sqrt{48} = \sqrt{9 \times 3} - \sqrt{25 \times 3} + \sqrt{16 \times 3}$
$$= 3\sqrt{3} - 5\sqrt{3} + 4\sqrt{3}$$
$$= 2\sqrt{3}.$$

(2) Rationalisation

The evaluation of a number such as $\dfrac{1}{\sqrt{2}}$ will be easier, and likely to be more accurate, if the fraction can be transformed so that we multiply by the surd and do not divide by it. This can be done by the following transformation:

$$\frac{1}{\sqrt{2}} = \frac{\sqrt{2}}{\sqrt{2} \times \sqrt{2}} = \frac{\sqrt{2}}{2}.$$

Similarly

$$\frac{\sqrt{5}}{2\sqrt{3}} = \frac{\sqrt{5} \times \sqrt{3}}{2\sqrt{3} \times \sqrt{3}} = \frac{\sqrt{15}}{2 \times 3} = \frac{\sqrt{15}}{6}.$$

By this transformation the denominator is changed from an irrational number to a rational one.

This is called rationalising the denominator.

If the denominator is a binomial expression the method is slightly more difficult. The procedure is indicated in the following examples:

Example 1: Rationalise the denominator of $\dfrac{1}{\sqrt{5} - \sqrt{2}}$.

Since $(a - b) \times (a + b) = a^2 - b^2$ and $(\sqrt{a} + \sqrt{b}) \times (\sqrt{a} - \sqrt{b}) = a - b$ (see above), then, if the denominator is multiplied by $\sqrt{5} + \sqrt{2}$, the surds will disappear from it.

Thus
$$\frac{1}{\sqrt{5} - \sqrt{2}} = \frac{\sqrt{5} + \sqrt{2}}{(\sqrt{5} - \sqrt{2})(\sqrt{5} + \sqrt{2})}$$
$$= \frac{\sqrt{5} + \sqrt{2}}{5 - 2} = \frac{\sqrt{5} + \sqrt{2}}{3}.$$

Example 2: Simplify $\dfrac{\sqrt{5} - 1}{\sqrt{5} + 1}$.

To rationalise the denominator it must be multiplied by $\sqrt{5} - 1$.

$$\therefore \quad \frac{\sqrt{5} - 1}{\sqrt{5} + 1} = \frac{(\sqrt{5} - 1)(\sqrt{5} - 1)}{(\sqrt{5} + 1)(\sqrt{5} - 1)} = \frac{(\sqrt{5} - 1)^2}{(\sqrt{5})^2 - (1)^2}$$
$$= \frac{5 - 2\sqrt{5} + 1}{5 - 1}$$
$$= \frac{6 - 2\sqrt{5}}{4} = \frac{3 - \sqrt{5}}{2}.$$

Exercise 69

1 Express the following as complete square roots, thus:

$$3\sqrt{7} = \sqrt{9 \times 7} = \sqrt{63}.$$

(1) $5\sqrt{6}$ (2) $12\sqrt{2}$
(3) $10\sqrt{5}$ (4) $4\sqrt{13}$
(5) $2a\sqrt{b}$ (6) $3x^2\sqrt{y}$

2 Express the following with the smallest possible number under the root sign, in each case:

(1) $\sqrt{800}$ (2) $\sqrt{320}$
(3) $\sqrt{108}$ (4) $\sqrt{5000}$
(5) $\sqrt{375}$ (6) $\sqrt{7200}$
(7) $\sqrt{24a^3b^2}$ (8) $\sqrt{18x^4y^3z}$
(9) $\sqrt{75a^3b^3c^3}$ (10) $\sqrt{1000p^2q}$
(11) $\sqrt{a^2b(a - b)^3}$ (12) $\sqrt{25xy^3(x - 2y)^2}$

3 Simplify:

(1) $\sqrt{3} \times \sqrt{15}$ (2) $\sqrt{14} \times \sqrt{7}$
(3) $\sqrt{32} \times \sqrt{24}$ (4) $2\sqrt{5} \times 5\sqrt{2}$
(5) $\sqrt{2}(2\sqrt{2} - 1)$ (6) $\sqrt{7}(\sqrt{14} + \sqrt{2})$
(7) $\sqrt{a + b} \times \sqrt{a^2 - b^2}$
(8) $\sqrt{3(x - 2y)} \times \sqrt{6(x^2 - 4y^2)}$

4 Multiply the following:

(1) $(\sqrt{2} - 1)(\sqrt{2} + 3)$
(2) $(2\sqrt{3} + \sqrt{2})(\sqrt{3} - 2\sqrt{2})$
(3) $(\sqrt{2} - 1)^2$ (4) $(2\sqrt{5} + \sqrt{3})^2$
(5) $(\sqrt{7} - 5\sqrt{2})^2$
(6) $(\sqrt{5} + \sqrt{3})(\sqrt{5} - \sqrt{3})$
(7) $(2\sqrt{10} - \sqrt{2})(2\sqrt{10} + \sqrt{2})$
(8) $(1 + 10\sqrt{3})^2$ (9) $(\sqrt{a} - 5)(\sqrt{a} + 5)$
(10) $(\sqrt{27} + \sqrt{6})(\sqrt{27} - \sqrt{6})$

5 Simplify the following by rationalising the denominator:

(1) $\dfrac{2}{\sqrt{3}}$ (2) $\dfrac{1}{\sqrt{5}}$

(3) $\dfrac{3}{2\sqrt{7}}$

(4) $\dfrac{1}{3\sqrt{10}}$

(5) $\dfrac{1}{\sqrt{32}}$

(6) $\dfrac{12}{\sqrt{20}}$

(7) $\dfrac{\sqrt{2}}{\sqrt{3}}$

(8) $\dfrac{2\sqrt{5}}{\sqrt{7}}$

(9) $\dfrac{1}{\sqrt{5}-1}$

(10) $\dfrac{3}{\sqrt{7}+\sqrt{2}}$

(11) $\dfrac{3}{2\sqrt{3}-\sqrt{2}}$

(12) $\dfrac{5}{2\sqrt{7}-3\sqrt{3}}$

(13) $\dfrac{\sqrt{3}+\sqrt{2}}{\sqrt{3}-\sqrt{2}}$

(14) $\dfrac{\sqrt{7}-\sqrt{5}}{\sqrt{7}+\sqrt{5}}$

(15) $\dfrac{\sqrt{7}}{3\sqrt{7}-1}$

(16) $\dfrac{1}{\sqrt{2}-1}+\dfrac{1}{\sqrt{2}+1}$

(17) $\dfrac{1}{\sqrt{5}+\sqrt{3}}-\dfrac{1}{\sqrt{5}-\sqrt{3}}$

(18) $\dfrac{4}{6-\sqrt{5}}$

24

Arithmetical and Geometrical Series

204 Meaning of a series

A series is a succession of numbers each of which is formed according to a definite law, which is the same throughout the series.

The ordinary numbers 1, 2, 3, 4, . . . constitute a series, each term of which is greater by unity than the one which immediately precedes it.

5, 9, 13, 17, . . . is a series, each term of which is greater by 4 than the one which immediately precedes it.

2, 4, 8, 16, . . . is a series in which each term is twice the one which immediately precedes it.

$1, \frac{1}{2}, \frac{1}{3}, \frac{1}{4}, \ldots$ is a series in which the terms are the reciprocals of 1, 2, 3, 4, . . .

$1^2, 2^2, 3^2, \ldots$
$1^3, 2^3, 3^3, \ldots$

are series, the construction of each of which is obvious.

205 The formation of a series

Series are of great importance in modern mathematics, but in this book it is possible to deal with only a few simple cases.

The two most important things to be known about a series are:

(1) The law of its formation. If this is known, it is possible to find any term in the series.

(2) The sum of a given number of terms of the series.

In this connection it is necessary to consider what is the nature of the sum when the number of terms is very great. If the series is one in which the terms increase numerically such as

$$2, 5, 8, 11, \ldots$$
or
$$1, 3, 9, 27, \ldots$$

it is clear that the more terms which are taken, the greater will be the sum. But if the series is one in which the terms decrease as the number of terms increases, such as $1, \frac{1}{3}, \frac{1}{9}, \frac{1}{27}, \ldots$, it is not always easy to discover what the sum will be when the number of terms is very great. This is a matter which will be considered later.

206   Arithmetic series

An arithmetic series, or arithmetic progression, is one in which each term is formed from that immediately preceding it by adding or subtracting a constant number.

The number thus added or subtracted is called the common difference of the series.

Examples

(1) 7, 13, 19, 25, . . . (common difference 6).
(2) 6, 4, 2, 0, -2, . . . (common difference -2).

In general if three numbers a, b, c, are in arithmetic progression (denoted by A.P.) then

$$b - a = c - b.$$

207   Any term in an arithmetic series

Let a = the first term of a series
   d = the common difference ($+$ve or $-$ve)
then the series can be written

$$a, a + d, a + 2d, a + 3d, \ldots$$

It is evident that the multiple of d which is added to a to produce any term is one less than the number of the term in the sequence.

Thus the fourth term is $a + (4 - 1)d = a + 3d$. Hence if the number of any term be denoted by n

then
$$\text{nth term} = a + (n - 1)d.$$

Examples

(1) In the series 7, 10, 13, . . . the common difference is 3.
  ∴ the tenth term is $7 + (10 - 1)3 = 34$.
  ∴ the nth term is $7 + (n - 1)3$.
(2) In the series $6, 2, -2, -6, \ldots d = -4$.
  ∴ the nth term $= 6 + (n - 1)(-4) = 6 - (n - 1)4$.
  ∴ the eighth term $= 6 + 7(-4) = 6 - 28 = -22$.

208 The sum of any number of terms of an arithmetic series

Using the following symbols in addition to those used previously:

Let n = number of terms whose sum is required
  s = the sum of n terms
  l = the last term

then by the previous formula

$$l - a + (n - 1)d.$$

Now $s = a + (a + d) + (a + 2d) + \ldots + (l - d) + l$.

Reversing the series

$$s = l + (l - d) + (l - 2d) + \ldots + (a + d) + a.$$

Adding the corresponding terms of the two sets, each pair gives $(a + l)$.

$$2s = (a + l) + (a + l) + (a + l) + \ldots + (a + l) + (a + l)$$
$$= (a + l) \times n, \text{ since there are n terms and } \therefore \text{ n pairs.}$$

∴ $$s = \frac{n(a + l)}{2}.$$

Since $l = a + (n - 1)d$, on substituting for l in the last result

$$s = \frac{n\{a + a + (n - 1)d\}}{2}.$$

∴ $$s = \frac{n}{2}\{2a + (n - 1)d\}.$$

This formula, like all other formulae, may be used not only to find s, but also any of the other numbers n, a, or d.

To find a and d offers no difficulty, but if n is required it will be seen that a quadratic equation will result. Since there are two roots to every quadratic equation, two values of n will always be found. In some cases only one root is admissible; in others both roots provide solutions.

For example, in a series involving negative terms such as

$$9 + 7 + 5 + 3 + 1 - 1 - 3$$

it is seen that the sum of 7 terms is the same as the sum of 3 terms. In other cases it will be evident that one of the roots is inadmissible.

209   Arithmetic mean

If three numbers are in arithmetic progression, the middle one is called the arithmetic mean of the other two.

Let a, b, c, be three numbers in A.P.

then by the definition of Section 206

$$b - a = c - b$$
$$2b = a + c.$$
$$\therefore \quad b = \frac{a + c}{2}.$$

It will be seen that the arithmetic mean of two numbers is the same as their average.

It is usual also to speak of inserting arithmetic means between two numbers, by which is meant that they, together with the two given, form a series of numbers in A.P.

Example: Insert three arithmetic means between 4 and 20.

If these be a, b, c, then 4, a, b, c, 20 are in A.P., five terms in all. Using $l = a + (n - 1)d$ for the fifth term 20
$$20 = 4 + (5 - 1)d, \text{ whence } d = 4.$$

∴ the five terms are 4, 8, 12, 16, 20.

210   Worked examples

Example 1: The sum of an A.P. of 8 terms is 90 and the first term is 6. What is the common difference?

Using $$s = \frac{n}{2} \{2a + (n - 1)d\}.$$

and substituting given values

$$90 = \tfrac{8}{2}\{2 \times 6) + (8 - 1)d\}$$
$$90 = 4(12 + 7d) = 48 + 28d.$$
$$\therefore \qquad 28d = 42$$
and $\qquad d = 1.5.$

Example 2: How many terms of the series 3, 6, 9, . . . must be taken so that their sum is 135?

Using $\qquad\qquad\qquad s = \dfrac{n}{2}\{2a + (n - 1)d\}$.

and substituting, $\qquad\qquad 135 = \dfrac{n}{2}\{6 + (n - 1)3\}$.

$\therefore \qquad\qquad\qquad 270 = n(3 + 3n)$
or $\qquad\qquad\qquad 270 = 3n + 3n^2.$
$\therefore \qquad\qquad n^2 + n - 90 = 0.$

Factorising $\quad (n - 9)(n + 10) = 0$
$\therefore \qquad\qquad\qquad n = 9 \text{ or } -10.$

The root -10 is inadmissible as having no meaning in this connection.

$\therefore$ the solution is $\qquad\qquad n = 9.$

Exercise 70

1 Write down the next three terms of the following series:

    (1) 5, 7.5, 10, . . .
    (2) 12, 8, 4, . . .
    (3) $(a + 3b), (a + b), (a - b)$
    (4) 2.7, 4, 5.3
    (5) $x - y, x, x + y$

2 Find the fifth and eighth terms of the series whose first term is 6, and common difference 1.5.

3 Find the 2pth term of the series whose first term is 6 and common difference 2.

4 Find the nth term of the series whose first term is $(x + 2)$ and common difference 3.

5 Find the twenty-fifth term of the series 0.6, 0.72, 0.84, . . .

6 The fourth term of an A.P. is 11 and the sixth term 17. Find the tenth term.

7 The fifth term of an A.P. is 11 and the ninth term is 7. Find the fourteenth term.

8 Which term of the series 2.3, 4.2, 6.1, . . . is 36.5?

9 Find the sums of the following series:

> (*a*) 15, 16.5, 18, . . . to ten terms.
> (*b*) 9, 7, 5, . . . to eight terms.
> (*c*) 0.8, 0.6, 0.4, . . . to nine terms.
> (*d*) $2\frac{2}{3}$, $3\frac{1}{2}$, $4\frac{1}{3}$, . . . to twenty-seven terms.

10 How many terms of the series 10, 12, 14, . . . must be taken so that the sum of the series is 252?

11 How many terms of the series 24, 20, 16, . . . must be taken so that the sum of the series is 80?

12 Find the thirtieth term and the sum of thirty terms of the series 4, 8, 12, . . .

13 A contractor agrees to sink a well 250 m deep at a cost of £2.70 for the first metre, £2.85 for the second metre, and an extra 15p for each additional metre. Find the cost of the last metre and the total cost.

14 A parent places in the savings bank 25p on his son's first birthday, 50p on his second, 75p on his third, and so on, increasing the amount by 25p on each birthday. How much will be saved up when the boy reaches his sixteenth birthday, the latter inclusive?

---

211 Harmonic progression

A series of numbers is said to be in harmonic progression (H.P.) if their reciprocals form a series in arithmetic progression.

> Thus the series 1, 3, 5, 7, . . . are in A.P.
>
> ∴ the series 1, $\frac{1}{3}$, $\frac{1}{5}$, $\frac{1}{7}$, . . . are in H.P.

This series is important in the theory of sound.

It is not possible to obtain a formula for the sum of n terms of an H.P., but many problems relating to such a series can be solved by using the corresponding arithmetic series.

Harmonic mean

– The harmonic mean of two numbers may be found as follows:

Let a and b be the numbers.
Let H be their harmonic mean,

> i.e., a, H, b are in H.P.

> Then $\frac{1}{a}$, $\frac{1}{H}$, $\frac{1}{b}$ are in A.P.

$$\therefore \qquad \frac{1}{H} = \frac{1}{2}\left(\frac{1}{a} + \frac{1}{b}\right) \text{ (Section 209)}$$

$$= \frac{a + b}{2ab}.$$

$$\therefore \qquad H = \frac{2ab}{a + b}.$$

212   Geometric series, or geometric progression

A geometric series is one in which the ratio of any term to that which immediately precedes it is constant for the whole series.

This ratio is called the common ratio of the series. It may be positive or negative. Thus each term of the series can be obtained by multiplying the term which precedes it by the common ratio.

Examples

(1) 1, 2, 4, 8, . . .   (common ratio 2).
(2) 1, $\frac{1}{2}$, $\frac{1}{4}$, $\frac{1}{8}$, . . .   (common ratio $\frac{1}{2}$).
(3) 2, -6, 18, -54, . . .   (common ratio -3).
(4) R, R^2, R^3, R^4, . . .   (common ratio R).

If three numbers a, b, c are in geometric progresion (G.P.) then $\frac{b}{a} = \frac{c}{b}$.

This is the test to apply in order to find if numbers are in G.P.

General form of a geometric series

Let a = 1st term.
Let r = common ratio.

Then the series is a, ar, ar^2, ar^3, . . .

213   Connection between a geometric series and an arithmetic series

In the geometric series

$$a, ar, ar^2, ar^3, \ldots$$

take logs of each term. We then get the series:

$$\log a, \log a + \log r, \log a + 2 \log r, \log a + 3 \log r, \ldots$$

This is an arithmetic series in which the first term is log a, and the common difference is log r.

∴ the logarithms of the terms of a G.P. form a series in A.P.

214   General term of a geometric series

Examining the series a, ar, ar^2, ar^3, . . . it will be seen that each term of the series is the product of a and a power of r the index of which is one less than the number of the term.

∴ if n = any term then the nth term = ar^{n-1}.

If r is negative, n − 1 being alternately odd and even, the terms will be alternately negative and positive, assuming a to be positive.

When n − 1 is even, n is odd, and the nth-term is + ve.

When n − 1 is odd, n is even, and the nth-term is − ve.

Worked examples

Example 1: Find the seventh term of the series

$$3, 6, 12, . . .$$

In this series r = 2, so using the formula

$$\text{nth term} = ar^{n-1}$$
$$\text{the seventh term} = ar^6 = 3 \times 2^6 = 3 \times 64$$
$$= 192.$$

Example 2: Find the eighth term of the series

$$2, -6, 18, -54, + . . .$$

For this series r = − 3.
Using ar^{n-1},

$$\text{eighth term} = 2 \times (-3)^{8-1} = 2 \times (-3)^7$$
$$= -4374.$$

Example 3: Find the fifth term of the series in which the first term is 100 and the common ratio 0.63.

Using ar^{n-1}, if x = the fifth term

No.	log.
0.63	$\bar{1}$.7993
	4
	$\bar{1}$.1972
100	2
15.75	1.1972

$$x = 100 \times (0.63)^4.$$
∴ $\log x = \log 100 + 4 \log 0.63$
$$= 1.1972$$
$$= \log 15.75.$$

∴ the fifth term is 15.75.

Example 4: The third term of a G.P. is 4.5 and the ninth is 16.2. Find the common ratio.

Using ar^{n-1},

$$\text{third term} = ar^2 = 4.5$$
$$\text{ninth term} = ar^8 = 16.2.$$

Dividing $\qquad ar^8 \div ar^2 = 16.2 \div 4.5.$

∴ $\qquad\qquad r^6 = 16.2 \div 4.5.$

Taking logs $\qquad 6 \log r = \log 16.2 - \log 4.5$
$$= 0.5563.$$

∴ $\log r = 0.5563 \div 6$
$$= 0.0927$$
$$= \log 1.238.$$

∴ $\qquad r = 1.238.$

No.	log.
16.2	1.2095
4.5	0.6532
	0.5563

215 Geometric mean

If three numbers are in G.P., the middle term is called the geometric mean of the other two.

Let a, b, c be three numbers in G.P.

Then by the definition of Section 212

$$\frac{b}{a} = \frac{c}{b}.$$

∴ $\qquad\qquad b^2 = ac,$

and $\qquad\qquad b = \sqrt{ac}.$

Exercise 71

1 Write down the next three terms of each of the following series:

    (*a*) 4, 10, 25
    (*b*) 16, 4, 1
    (*c*) 16, −24, 36
    (*d*) 0.3, 0.03, 0.003
    (*e*) 3, 0.45, 0.0675

2 Find the seventh term of the series 5, 10, 20, . . .
3 Find the seventh term of the series 6, −4, $2\frac{2}{3}$, . . .
4 Find the fifth term of the series 1.1, 1.21, 1.331, . . .
5 Find the sixth term of the series −0.5, 0.15, −0.045, . . .

6 Write down the 2nth and the $(2n + 1)$th terms of the series:

(1) $a, ar, ar^2, \ldots$
(2) $a, -ar, ar^2, \ldots$

7 The first term of a G.P. is 1.05 and the sixth term is 1.3401. Find the common ratio.

8 The fifth term of a G.P. is 1.2166 and the seventh term is 1.3159. Find the common ratio.

9 Find the geometric mean in each of the following cases:

(1) 3 and 5
(2) 4.2 and 3.6

10 Insert two geometric means between 5 and 13.72.

11 A man was appointed to a post at a salary of £1000 a year with an increase each year of 10 per cent of his salary for the previous year. How much does he receive during his fifth year?

12 The expenses of a company are £200 000 a year. It is decided that each year they shall be reduced by 5 per cent of those for the preceding year. What will be the expenses during the fourth year, the first reduction taking place during the first year?

13 In a geometric series the first term is unity and the fifth term is 1.170 approx. Find the common ratio.

14 Insert three terms in geometric progression between 5 and 80.

216 The sum of n terms of a geometric series

In addition to the symbols employed above,

Let S_n represent the sum of n terms of a G.P.

then
$$S_n = a + ar + ar^2 + \ldots + ar^{n-2} + ar^{n-1} \quad (1)$$

Multiply both sides by r,

then
$$rS_n = ar + ar^2 + ar^3 \ldots + ar^{n-1} + ar^n \quad (2)$$

Subtracting (1) from (2),

$$rS_n - S_n = ar^n - a$$

or
$$S_n(r - 1) = a(r^n - 1).$$

∴
$$S_n = \frac{a(r^n - 1)}{r - 1} \quad (A)$$

If (2) be subtracted from (1) above, the formula becomes:

$$S_n = \frac{a(1 - r^n)}{1 - r} \qquad \text{(B)}$$

If $r > 1$ and positive, form (A) should be used.
If $r < 1$ or negative, form (B) should be used.

Note: It will often be necessary to employ logarithms to evaluate r^n. They cannot of course be employed to evaluate the whole formula.

217 Worked examples

Example 1: Find the sum of seven terms of the series

$$2, 3, 4.5, \ldots$$

$$r = \tfrac{3}{2} = 1.5.$$

Using
$$S = \frac{a(r^n - 1)}{r - 1},$$

and substituting
$$S = \frac{2(1.5^7 - 1)}{1.5 - 1}$$

logs may be used to find 1.5^7.

Let
$$x = 1.5^7,$$

then
$$\begin{aligned}
\log x &= 7 \log 1.5 \\
&= 7 \times 0.1761 \\
&= 1.2327 \\
&= \log 17.09
\end{aligned}$$

$\therefore \qquad 1.5^7 = 17.09$ (approx.)

$$\begin{aligned}
\therefore \qquad S &= \frac{2(17.09 - 1)}{0.5} \\
&= 4 \times 16.09 \\
&= 64.36 = 64.4 \text{ approx.}
\end{aligned}$$

Example 2: Find the sum of seven terms of the series

$$4, -8, 16, \ldots$$

$$r = -2.$$

Using
$$S = \frac{a(1 - r^n)}{1 - r},$$

and substituting
$$S = \frac{4\{1 - (-2)^7\}}{1 - (-2)} = \frac{4(1 + 128)}{3}$$

$$= \tfrac{4}{3} \times 129 = 172.$$

Exercise 72

1 Find the sums of the following series:

  (*a*) 1.5, 3, 6, . . . to six terms.
  (*b*) 30, -15, $7\frac{1}{2}$, . . . to eight terms.
  (*c*) $\frac{1}{2}$, $-\frac{1}{4}$, $\frac{1}{8}$, . . . to six terms.

2 Find the sum of the first six terms of the series

$$5, 2.5, 1.25, \ldots$$

3 Find the sum of the first six terms of the series

$$1 - \tfrac{2}{3} + \tfrac{4}{9}, \ldots$$

4 Find the sum of the first six terms of the series

$$1 + 1.4 + 1.96 + \ldots$$

5 Find the sum of the first twelve terms of the series

$$4 + 5 + 6.25 + \ldots$$

6 If the first and third terms of a G.P. are 3 and 12, find the sum of eight terms.

7 If the third and fourth terms of a G.P. are $\frac{1}{9}$ and $\frac{1}{3}$, respectively, find the eighth term and the sum of eight terms.

8 Find the sum of $20 + 18 + 16.2 + \ldots$ to six terms.

218   Increasing series

When the common ratio of a geometric series is numerically greater than unity, as in the series

$$1, 2, 4, 8, \ldots$$
$$2.5, 7.5, 22.5, \ldots$$

the terms increase in magnitude. The sum of n terms increases as n increases. If the number of terms increases without limit – that is, n is greater than any number we may select, however great – then the sum of these terms will also increase without limit, i.e. it will become infinitely great, or, to use the mathematical term, approach 'infinity', which is denoted by the symbol ∞.

We may say, then, that as n, the number of terms, approaches infinity, S_n, the sum of these terms, also approaches infinity. This may be expressed by the following notation.

If $\qquad n \to \infty$, then $S_n \to \infty$.

219 Decreasing series

If, however, the common ratio is numerically less than unity, as in the following series,

$$\tfrac{1}{3}, \tfrac{1}{6}, \tfrac{1}{12}, \ldots$$
$$0.2, 0.02, 0.002, \ldots$$

then as the number of terms increases, the terms themselves decrease. Using the terms employed above, we may say that, as n increases without limit, the terms themselves decrease without limit, and ultimately become indefinitely small.

We cannot say, however, that the sum of these terms increases without limit, as n increases without limit. That is a matter for further investigation.

220 Recurring decimals

There is an example, arising from arithmetic, which will assist in coming to conclusions on this important question, viz. that of a recurring decimal. We know that

$$\tfrac{1}{9} = 0.1111 \ldots$$

in which 1 recurs without limit.

The decimal is in effect the geometric series

$$\frac{1}{10} + \frac{1}{10^2} + \frac{1}{10^3} + \ldots$$

in which there is no limit to the number of terms. It is an example of what is called an infinite series. But we know that the sum of all these terms, no matter how many are taken, is ultimately equal to the finite fraction $\tfrac{1}{9}$.

If we find the sum of finite numbers of terms, we get:

$$S_1 = \frac{1}{10}.$$

$$S_2 = \frac{1}{10} + \frac{1}{10^2} = \frac{11}{10^2}.$$

$$S_3 = \frac{1}{10} + \frac{1}{10^2} + \frac{1}{10^3} = \frac{111}{10^3} \text{ and so on.}$$

The difference between $\frac{1}{9}$ and the sums of these is:

$$\frac{1}{9} - S_1 = \frac{1}{90}$$
$$\frac{1}{9} - S_2 = \frac{1}{900}$$
$$\frac{1}{9} - S_3 = \frac{1}{9000}$$

and in general, finding the sum of n terms by using the formula

$$S_n = \frac{a(1 - r^n)}{1 - r}$$

we get

$$S_n = \frac{\dfrac{1}{10}\left(1 - \dfrac{1}{10^n}\right)}{1 - \dfrac{1}{10}}$$

$$= \frac{1}{9}\left(1 - \frac{1}{10^n}\right) = \frac{1}{9} - \frac{1}{9 \times 10^n}.$$

Examining these results, it is seen that the difference between $\frac{1}{9}$ and the various sums, $S_1, S_2, S_3, \ldots S_n$ decreases as n increases. In general, the difference between $\frac{1}{9}$ and the sum of n terms is

$$\frac{1}{9 \times 10^n}.$$

As n increases without limit, this difference decreases without limit – i.e., it tends to become zero – and the sum approaches to equality with $\frac{1}{9}$. It can never be greater than $\frac{1}{9}$.

Using the previous notation, we can express the result thus

$$\text{as } n \to \infty, S_n \to \tfrac{1}{9}.$$

There is thus a limit to which S_n approaches and which it cannot exceed.

221   A geometrical illustration

The approach of the sum of a geometric series to a limit may be illustrated by a graphical representation of the series

$$\frac{1}{2} + \frac{1}{4} + \frac{1}{8} + \frac{1}{16} + \ldots$$

or

$$\frac{1}{2} + \frac{1}{2^2} + \frac{1}{2^3} + \frac{1}{2^4} + \ldots$$

Let the rectangle ABCD (Fig. 52) represent a unit of area.

Fig. 52.

Let E be the midpoint of AB and draw EF perpendicular to DC. Then rectangle AEFD represents $\frac{1}{2}$ of a unit.

Bisecting the rectangle EBCF by GH, then rectangle EGHF represents $\frac{1}{4}$ or $\frac{1}{2^2}$ of a unit.

Continuing the process of bisecting the rectangle left over after each bisection, we get a series of rectangles whose areas represent the terms of the above series. These rectangles diminish in area as we represent more and more terms of the series in this way.

The rectangle AKLD represents the sum of the four terms of the series

$$\frac{1}{2} + \frac{1}{2^2} + \frac{1}{2^3} + \frac{1}{2^4}.$$

As more divisions are made and the sum of more terms represented by a rectangle, this rectangle approaches nearer to the area of the whole rectangle – i.e., 1 – but can never exceed it. Consequently 1 is a limit which the sum of the series approaches as the number of terms in increased without limit, but which it can never exceed, no matter how many terms are taken.

If the series $\frac{1}{2} + \frac{1}{2^2} + \frac{1}{2^3} + \ldots$ be summed by using the formula

$$S_n = \frac{a(1 - r^n)}{1 - r},$$

we get

$$S_n = \frac{\frac{1}{2}\{(1 - (\frac{1}{2})^n\}}{1 - \frac{1}{2}} = \frac{\frac{1}{2}\{1 - (\frac{1}{2})^n\}}{\frac{1}{2}},$$

$$\therefore \qquad S_n = 1 - (\tfrac{1}{2})^n.$$

Examining this result, we see that $(\frac{1}{2})^n$ decreases as n increases. If n be increased without limit then $(\frac{1}{2})^n$ decreases without limit – i.e., it approaches zero.

$\therefore$ we can say that $S_n \to 1$ as $n \to \infty$.

222 The sum to infinity

The above suggests the general treatment of this question.

Using
$$S_n = \frac{a(1 - r^n)}{1 - r}$$

i.e.,
$$S_n = \frac{a - ar^n}{1 - r}$$

we have
$$S_n = \frac{a}{1 - r} - a \cdot \frac{r^n}{1 - r}.$$

Considering the term $a \cdot \dfrac{r^n}{1 - r}$, if r is a proper fraction (i.e., it lies between $+1$ and -1) then r^n diminishes as n increases or, with the previous notation,

$$\text{as } n \to \infty, r^n \to 0 \text{ and } a \cdot \frac{r^n}{1 - r} \to 0.$$

Thus the right hand side approaches $\dfrac{a}{1 - r}$ as a limit.

This is the 'limiting sum' of the series and it is called the 'sum to infinity'.

If it be represented by S_∞

then
$$S_\infty = \frac{a}{1 - r}.$$

223 Worked examples

Example 1: Sum to inifinity the series
$$2 + \tfrac{1}{2} + \tfrac{1}{8} + \ldots$$

Here $a = 2, r = \tfrac{1}{4}$.

$$\therefore \quad S_\infty = \frac{2}{1 - \tfrac{1}{4}} = 2 \div \tfrac{3}{4}$$
$$= 2\tfrac{2}{3}.$$

Example 2: Find the sum to infinity of the series
$$5 - 1 + \tfrac{1}{5} - \ldots$$

Here $a = 5, r = -\tfrac{1}{5}$.

$$\therefore \quad S_\infty = \frac{5}{1 - (-\tfrac{1}{5})} = 5 \div 1\tfrac{1}{5}$$
$$= 4\tfrac{1}{6}.$$

Exercise 73

1 To what limits will the following series tend as the number of terms increases indefinitely?

(a) $\frac{1}{2} + \frac{1}{8} + \frac{1}{32} + \ldots$
(b) $\frac{1}{4} + \frac{1}{16} + \frac{1}{64} + \ldots$
(c) $\frac{1}{2} + \frac{1}{4} + \frac{1}{8} + \frac{1}{16} + \frac{1}{32} + \ldots$

2 To what limits will the following series tend as the number of terms increases indefinitely?

(a) $0.1 + 0.001 + 0.000\ 01 + \ldots$
(b) $0.06 + 0.006 + 0.000\ 006 + \ldots$
(c) $0.16 + 0.0016 + 0.000\ 016 + \ldots$

What is the connection between these series?

3 To what limit does an infinite number of terms in the following series tend?

$$1 - \frac{1}{2} + \frac{1}{4} - \frac{1}{8} + \ldots$$

Show the connection with the series in question 1.

4 Show that the sum of n terms of the series

$$1 + \frac{1}{3} + \frac{1}{9} + \ldots$$

is

$$\frac{3}{2}\{1 - (\frac{1}{3})^n\}.$$

Hence show what limit this series approaches.

5 Find the sum of n terms of the series

$$1 + \frac{a}{b} + \frac{a^2}{b^2} \ldots$$

when

$$\frac{a}{b} < 1.$$

Hence find the limit approached by the series as the number of terms becomes infinitely great.

6 Find the limiting sum, or the sum to infinity of the following series:

(a) $\frac{8}{3} + \frac{4}{9} + \frac{2}{27} + \ldots$
(b) $5 - 1 + \frac{1}{5} - \ldots$
(c) $9 - 6 + 4 - \ldots$

7 Find the sum to infinity of the series

(a) $1 + \dfrac{1}{1.04} + \dfrac{1}{1.04^2} + \ldots$

(b) $\sqrt{2} + 1 + \dfrac{1}{\sqrt{2}} + \ldots$

8 The sum to infinity of a series is 15, and the first term is 3. Find the common ratio.

9 A superball is dropped from a height of 10 m. At each rebound it rises to a height which is 0.9 of the height from which it has just fallen. What is the total distance through which the ball will have moved before it finally comes to rest?

10 The yearly output of a silver mine is found to be decreasing by 25 per cent of its previous year's output. If in a certain year its output was £25 000 what could be reckoned as its total future output?

224 Simple and compound interest

The accumulation of money when put to interest furnishes examples of arithmetic and geometric series. When money is put out at simple interest, the interest is payable for each year, but is not added to the principal.

For example the interest payable on £100 at 5 per cent for 1, 2, 3 . . . years will be £5, £10, £15 . . . , these sums forming a series in A.P. and the interest varies directly as the time.

But if money is lent at compound interest, the interest is added each year to the principal, and for the following year the interest is calculated on their sum.

Suppose £1 to be invested at 5 per cent compound interest. Then the interest for the first year is £$\frac{5}{100}$ or £0.05.

∴ the amount at the end of the year is £1.05 and the amount at the end of the year of £P is £P × 1.05.

Consequently the ratio of the amount at the end of a year to that at the beginning is always 1.05.

This corresponds to the ratio of a geometric series.

∴ amount at the end of the second year is

$$P \times 1.05 \times 1.05 = P \times 1.05^2.$$

∴ amount at the end of the third year is

$$P \times 1.05^2 \times 1.05 = P \times 1.05^3.$$

∴ amount at the end of the fourth year is

$$P \times 1.05^3 \times 1.05 = P \times 1.05^4.$$

∴ amount at the end of the nth year

$$= P \times 1.05^n.$$

These amounts at the end of successive years, viz.:

$$P \times 1.05, P \times 1.05^2, P \times 105^3 \ldots$$

constitute a geometric series.

Let M = the amount at the end of n years

then $\qquad\qquad$ $M = PR^n$.

In this formula, as we have seen in other cases, any one of the four quantities may be the subject of the formula.

Thus $\qquad\qquad$ $P = \dfrac{M}{R^n}$.

This enables us to find the sum of money which will produce £M in n years.

Again $\qquad\qquad$ $R^n = \dfrac{M}{P}$.

∴ $\qquad\qquad$ $R = \sqrt[n]{\dfrac{M}{P}}$,

from which the rate of interest can be discovered.

Again $\qquad\qquad$ $n \log R = \log M - \log P.$

∴ $\qquad\qquad$ $n = \dfrac{\log M - \log P}{\log R}$

whence the time taken for P to amount to M is found.

225 Accumulated value of periodical payments

Suppose that £P is invested each year for 10 years at 5 per cent C.I., each investment being made at the beginning of a year.

Using the above formula:

The first £P at the end of 10 years amounts to

$$P \times 1.05^{10}.$$

The second £P at the end of 9 years amounts to

$$P \times 1.05^9.$$

The third £P at the end of 8 years amounts to

$$P \times 1.05^8.$$

And finally, the last £P invested bears interest for 1 year and amounts to

$$P \times 1.05.$$

Then the accumulated value of the investments amounts to

$$P \times 1.05^{10} + P \times 1.05^9 + P \times 1.05^8 \ldots + P \times 1.05,$$

or, reversing the series,

$$P \times 1.05 + P \times 1.05^2 + P \times 1.05^3 \ldots + P \times 1.05^{10},$$

This is a geometric series and using the form

$$S_n = \frac{a(r^n - 1)}{r - 1}$$

the accumulated value of the investments is

$$\frac{1.05(1.05^{10} - 1)}{1.05 - 1} \times P.$$

226 Annuities

An annuity is a series of equal annual payments extending over a specified number of years, or for the life of the annuitant.

A ground rent is a similar financial transaction, the holder of the freehold receiving an annual payment, called ground rent, for the number of years specified in the lease. Ground rents and annuities are constantly being bought and sold, and the method of calculating the amount to be paid by the purchaser can be determined by means of the above results. This amount will depend upon the rate of interest which the purchaser expects to receive on his investment.

Let the rate of interest expected be 4 per cent.

The price is obtained by finding the present value of each of the payments, as follows:

From the formula $M = PR^n$

we get as shown $P = \dfrac{M}{R^n}.$

P is the amount which produces M in n years at the given rate per cent; it is called the present value of M due in n years.

If £A be the annual payment and P be its present value, then for the first payment due in 1 year

$$P = \frac{A}{R} = \frac{A}{1.04}$$

when the rate per cent is 4.

For the second payment

$$P = \frac{A}{1.04^2} .$$

For the third payment

$$P = \frac{A}{1.04^3} ,$$

and so on.

∴ the total present value

$$= \frac{A}{1.04} + \frac{A}{1.04^2} + \frac{A}{1.04^3} + \cdots$$

$$= A \left\{ \frac{1}{1.04} + \frac{1}{1.04^2} + \frac{1}{1.04^3} + \cdots \right\}$$

$$= A \left\{ \frac{\frac{1}{1.04} \left[1 - \left(\frac{1}{1.04} \right)^n \right]}{1 - \frac{1}{1.04}} \right\}$$

$$= \frac{A}{0.04} \left\{ 1 - \frac{1}{1.04^n} \right\}$$

This can then be evaluated for any value of A.

The terms of the above series decrease and if the ground rent is a perpetual one, or the lease is a very long one, the present value becomes the sum to infinity of the above series – i.e.,

Present value

$$= A \left\{ \frac{\frac{1}{1.04}}{1 - \frac{1}{1.04}} \right\}$$

$$= A \left\{ \frac{\frac{1}{1.04}}{\frac{1.04 - 1}{1.04}} \right\} = A \frac{1}{0.04} = A \times \frac{100}{4} = A \times 25.$$

The ground rent is then said to be worth 25 years purchase. It is always found by dividing 100 by the rate per cent.

Exercise 74

1 If £100 be invested at the beginning of each year for 10 years at 3 per cent C.I., find the accumulated value a year after the last amount is invested.

2 An annuity of £600 a year is allowed to accumulate at 3 per cent C.I. for 8 years. What was the total amount at the end.

3 A man saves £25 every half-year and invests it at C.I. at $4\frac{1}{2}$ per cent. What will be the amount of his savings in 8 years if the last amount saved bears interest for 6 months?

4 Find the present value of an annuity of £300 for 10 years, reckoning C.I. at 4 per cent, the first payment being due one year after purchase.

5 What should be the purchase price of an annuity of £500 for 8 years, reckoning C.I. at $3\frac{1}{2}$ per cent?

6 A man wished to endow in perpetuity an institution with a yearly sum of £200. If C.I. be reckoned at 4 per cent, what amount will be needed for it?

7 A man retires at 65, when the expectation of life is 10.34 years, with a pension of £200. What single payment would be the equivalent of this, reckoning C.I. at 4 per cent?

8 A pension of £6000 per annum was awarded to Nelson and his heirs for ever. If this be commuted into a single payment, what should that be, reckoning C.I. at $2\frac{1}{2}$ per cent?

Appendix

The following brief statement of permutations, combinations and the binomial theorem, and a note on the roots of a quadratic equation and set notation are given for the benefit of students who may need to use them in the differential calculus or other branches of more advanced mathematics.

Permutations and Combinations

1   Permutations

Consider the following example:

A party of 6 people arrived at a theatre and obtained 4 seats together and 2 separate. In how many different ways could the 4 seats in a row be filled if there are no restrictions as to where any of the 6 may sit?

Consider the first seat. Since any one of the 6 people may sit in it, it can be filled in 6 different ways. With each of these 6 ways, the second seat can be filled in 5 different ways, since 5 people are left to choose from.

∴ there are (6 × 5) different ways of filling the first two seats.

With each of the 6 × 5 or 30 ways of filling the first two seats, there are 4 ways of filling the third seat, since 4 people are left to choose from.

∴ there are (6 × 5 × 4) different ways of filling the first three seats.

Similarly the fourth seat can be filled in 3 ways and

∴ there are (6 × 5 × 4 × 3) different ways of filling the 4 seats – i.e., 360 ways.

Arrangements of a number of different objects in a row are

called permutations and the above problem was that of the permutation of 6 things 4 at a time.

This is expressed by a special notation – viz., $^6P_4$ or $_6P_4$.

It will be seen that if the people mentioned had 6 seats together, the number of permutations or arrangements in these seats would be $6 \times 5 \times 4 \times 3 \times 2 \times 1$. This product of all the integral numbers from 1 to 6 inclusive is called factorial 6 and is expressed by the $\lfloor6$ or 6 !

In general, the product of the integral numbers from 1 to n inclusive is denoted by $\lfloor n$ or n !

Thus $\lfloor n = n(n - 1)(n - 2) \ldots 3, 2, 1$ and is called factorial n.

2   Permutations of n things r at a time, or ^nP_r

This is the general treatment of the above special case and the method adopted to find the formula is the same.

There are r places to be filled and n different things to choose from.

The 1st place can be filled in n ways
The 2nd place can then be filled in $(n - 1)$ ways

since with each of the n ways of filling the 1st place, each of the $(n - 1)$ ways of filling the second can be associated.

∴ there are $n(n - 1)$ ways of filling the first two places.

Similarly there are

$n(n - 1)(n - 2)$ ways of filling the three places,
and $n(n - 1)(n - 2)(n - 3)$ ways of filling the four places.

∴ by inspection there are

$n(n - 1)(n - 2)(n - 3) \ldots \{n - (r - 1)\}$ ways of filling the r places.

∴ $^nP_r = n(n - 1)(n - 2)(n - 3) \ldots (n - r + 1)$.

If the n things are all arranged among themselves, then the last factor becomes $(n - n + 1)$ or 1.

∴ $^nP_n = \lfloor n$.

3   Combinations

The problem solved above – viz., the number of different ways of filling up 4 seats by 6 people – might have been approached in another way.

(1) We could find the number of different sets or groups of 4 that could be formed from 6 people.
(2) Each group could then be arranged in the seats in $\lfloor 4$ ways.

The product of these two numbers must give the total number of ways of filling the 4 seats – i.e., the permutations of 6 things 4 at a time. The difficulty at present is that of finding the number of groups.

Let x = the number of groups

then, by the above reasoning,

$$x = \lfloor 4 = {}^6P_4.$$
$$\therefore \qquad x = {}^6P_4 \div \lfloor 4.$$

Thus we can find the number of groups, when we know the number of permutations.

Such groups are called combinations and a notation similar to that used for permutation is employed to express them.

Thus a number of groups or combinations of 6 different things 4 at a time is denoted by ${}^6C_4$, and in general, the number of combinations of n things r at a time is denoted by ${}^nC_r$ or ${}_nC_r$.

4 The Combinations of n things r at a time

With the same reasoning as that employed above we can deduce that

$${}^nC_r = {}^nP_r \div \lfloor r$$

or
$$\qquad {}^nC_r = \frac{n(n-1)(n-2)\ldots(n-r+1)}{\lfloor r}.$$

Thus
$$\qquad {}^nC_2 = \frac{n(n-1)}{\lfloor 2}$$

$$\qquad {}^nC_3 = \frac{n(n-1)(n-2)}{\lfloor 3}$$

$$\qquad {}^{10}C_4 = \frac{10 \times 9 \times 8 \times 7}{1 \times 2 \times 3 \times 4} = 210$$

$$\qquad {}^8C_3 = \frac{8 \times 7 \times 6}{1 \times 2 \times 3} = 56.$$

The Binomial Theorem

5   Products of binomial factors

It was shown in Section 91 that

$$(x + a)(x + b) = x^2 + x(a + b) + ab.$$

Employing the methods used in the chapter we can show that
$$(x + a)(x + b)(x + c) = x^3 + x^2(a + b + c) + x(ab + bc + ca) + abc.$$

It should be noted that:

(1) The expression is arranged in descending powers of x.
(2) The coefficients of these powers, after the first, are the sets formed in every way of the letters a, b, c,

   (*a*) one at a time, (*b*) 2 at a time, (*c*) 3 at a time.

From the way in which this product is formed we can deduce the product of

$$(x + a)(x + b)(x + c)(x + d).$$

Arranging the powers of x in descending order, the coefficients of these powers will be

x^4 – unity.
x^3 – sum of the letters one at a time, i.e.,

$$(a + b + c + d).$$

x^2 – sum of the letters two at a time, i.e.,

$$(ab + bc + ad + bd + cd + ac).$$

x – sum of the letters three at a time, i.e.,

$$(abc + bcd + acd + abd).$$

Term independent of x is abcd.

$\therefore$ the full product is
$$x^4 + x^3(a + b + c + d) + x^2(ab + bc + cd + ad + bd + ac) + x(abc + bcd + acd + abd) + abcd.$$

We have seen from Section 4, that the number of ways of

(1) Grouping 4 letters 1 at a time is $^4C_1 = \dfrac{4}{1} = 4$.

(2) Grouping 4 letters 2 at a time is $^4C_2 = \dfrac{4 \cdot 3}{1 \cdot 2} = 6$.

(3) Grouping 4 letters 3 at a time is $^4C_3 = \dfrac{4.3.2}{1.2.3} = 4$.

(4) Grouping 4 letters 4 at a time is $^4C_4 = 1$.

In the above factors let $b = c = d = a$
then the left side is $(x + a)^4$.
In the expansion of it:

The coefficient of x^3 is $(a + a + a + a) = 4a$.
The coefficient of x^2 is $6a^2$.
The coefficient of x is $4a^3$.

The last term is a^4.

$\therefore \ (x + a)^4 = x^4 + 4x^3a + 6x^2a^2 + 4xa^3 + a^4$
or $\quad (x + a)^4 = x^4 + {}^4C_1x^3a + {}^4C_2x^2a^2 + {}^4C_3xa^3 + a^4$.

By a similar process we can obtain the expansion

$(x + a)^5 = x^5 + {}^5C_1x^4a + {}^5C_2x^3a^2 + {}^5C_3x^2a^3 + {}^5C_4xa^4 + a^5$.

From a consideration of these results we can deduce the general case – viz.,

$$(x + a)^n = x^n + {}^nC_1x^{n-1}a + {}^nC_2x^{n-2}a^2 + {}^nC_3x^{n-3}a^3 +$$
$$\dots + a^6$$

or

$$(x + a)^n = x^n + nx^{n-1}a + \frac{n(n - 1)}{\underline{|2}}\,x^{n-2}a^2$$
$$+ \frac{n(n - 1)(n - 2)}{\underline{|3}}\,x^{n-3}a^3 + \dots + a^n.$$

This is called the binomial theorem.

The above reasoning is independent of the values of x and a. It will therefore hold if a be replaced by $-a$. Then

$$(x - a)^n = x^n + nx^{n-1}(-a) + \frac{n(n - 1)}{1.2}\,x^{n-2}(-a)^2 +$$
$$\dots + (-a)^n.$$

Since odd powers of $(-a)$ are negative and even powers are positive, the terms will be alternately $+$ve and $-$ve.

$$\therefore \ (x - a)^n = x^n - nx^{n-1}a + \frac{n(n - 1)}{1.2}\,.x^{n-2}a^2$$
$$- \frac{n(n - 1)(n - 2)}{1.2.3}\,x^{n-3}a^3 + \dots + (-a)^n.$$

In the previous results let x = 1.

Then

$$(1 + a)^n = 1 + na + \frac{n(n - 1)}{1.2} a^2 + \frac{n(n - 1)(n - 2)}{1.2.3} a^3 +$$

$$\ldots + a^n$$

and

$$(1 - a)^n = 1 - na + \frac{n(n - 1)}{1.2} a^2 - \frac{n(n - 1)(n - 2)}{1.2.3} a^3 +$$

$$\ldots + (-a)^n.$$

Every binomial expression can be reduced to one of these forms as follows:

$$(x + a)^n = \left\{x\left(1 + \frac{a}{x}\right)\right\}^n = x^n\left(1 + \frac{a}{x}\right)^n$$

$$= x^n\left\{1 + n.\frac{a}{x} + \frac{n(n - 1)}{1.2}\left(\frac{a}{x}\right)^2 + \ldots + \left(\frac{a}{x}\right)^n\right\}$$

Similarly,

$$(x - a)^n = x^n\left(1 - \frac{a}{x}\right)^n.$$

A complete proof of the binomial theorem requires a more advanced knowledge of algebra than is provided in this book. The demonstration given above assumes that n is a positive integer. In using the theorem later the important question will arise:

Is the binomial theorem true for fractional and negative indices?

It can be proved that the form holds for all values of n. For example of $n = \frac{1}{2}$.

$$(1 + x)^{\frac{1}{2}} = 1 + \frac{1}{2}x + \frac{\frac{1}{2}(\frac{1}{2} - 1)}{1.2} x^2 + \frac{\frac{1}{2}(\frac{1}{2} - 1)(\frac{1}{2} - 2)}{1.2.3} x^3 + \ldots$$

When n is a positive integer, a term in the series will ultimately be reached when one of the factors

$$n(n - 1)(n - 2) \ldots$$

will become (n - n) and will vanish, as will all succeeding products.

The number of terms therefore is finite and will clearly be n + 1.

Consequently when a value is assigned to x in $(1 + x)^n$ the sum of the series will be a finite number and the series is said to be convergent.

But if n is fractional or negative, none of the factors $n(n - 1)$ $(n - 2) \ldots$ will vanish, and the number of terms is infinite. If the sum increases without limit as n increases the series is said to be divergent. But it can be shown that if x, in $(1 + x)^n$, is numerically less than unity, the sum of the series will approach a limit, as is the case with certain geometric series (see Section 222). Subject to this condition, the series is always convergent.

The Roots of a Quadratic Equation

Writing the general quadratic equation

$$ax^2 + bx + c = 0$$

in the form

$$x^2 + \frac{b}{a}x + \frac{c}{a} = 0 \qquad \text{(A)}$$

the solution is $\quad x = \dfrac{-b \pm \sqrt{b^2 - 4ac}}{2a}$. (Section 143)

Let α and β be the two roots

then $\quad \alpha + \beta = \dfrac{-b + \sqrt{b^2 - 4ac}}{2a} + \dfrac{-b - \sqrt{b^2 - 4ac}}{2a}$

$$= -\frac{2b}{2a} = -\frac{b}{a}.$$

$$= \text{coeff. of x in equation (A).}$$

Again $\quad \alpha\beta = \dfrac{-b + \sqrt{b^2 - 4ac}}{2a} \times \dfrac{-b - \sqrt{b^2 - 4ac}}{2a}$

$$= \frac{(-b)^2 - (\sqrt{b^2 - 4ac})^2}{4a^2} \quad \text{(Section 97)}$$

$$= \frac{c}{a}$$

$$= \text{term independent of x in (A).}$$

Summarising, $\quad \alpha + \beta = -\dfrac{b}{a}$

$$\alpha\beta = \frac{c}{a} \quad \text{(cf. Section 91)}$$

Nature of the roots of a quadratic equation

If $(b^2 - 4ac)$ is negative, the square root has no arithmetical meaning. It is customary to speak of such a root as imaginary, while the square root of a positive number is called real.

Hence, for the roots of a quadratic,

(1) If $b^2 > 4ac$, the roots are real and different.
(2) If $b^2 = 4ac$ the roots are equal.
(3) If $b^2 < 4ac$ the roots are imaginary.

Such a number as $\sqrt{-p}$ can be written

$$\sqrt{p \times (-1)} = \sqrt{p} \times \sqrt{-1}.$$

The number $\sqrt{-1}$ is usually denoted by i.
Hence, as an example, $\sqrt{-9}$ can be written $\pm \sqrt{9 \times -1}$

$$= \pm 3i.$$

Example: The roots of the equation $x^2 + 2x + 5 = 0$ are given by

$$x = \frac{-2 \pm \sqrt{4 - 20}}{2}$$

$$= \frac{-2 \pm \sqrt{-16}}{2}$$

$$= \frac{-2 \pm 4i}{2}$$

$$= -1 \pm 2i.$$

Note: It will be found that the graph of $x^2 + 2x + 5$ does not cut the x axis, i.e., it does not equal zero for any real value of x.

Sets and Set Algebra

1   Set notation

In recent years an attempt has been made to provide a common language to unify many mathematical ideas. The language of sets is a helpful way of doing this as well as being an algebra in its own right. Many of the ideas are not really new at all but simply represented in a much tighter framework.

A set is simply a collection of *distinguishable* items which may or may not have any link between them. For exmaple we can talk about the set of people who live in a particular house, the set of vowels in the English alphabet (a, e, i, o and u), or the set of objects which happened to be in my pocket when I wrote this page, namely a paper clip, a two penny piece and a handkerchief. To make it clear what we are talking about the items are listed in squiggly brackets { . . . }.

The set of vowels is: {a, e, i, o, u}
The set in my pocket is: {a paper clip, a two penny piece, a handkerchief}

A set can be defined by listing the items which belong to it, as in the examples above, or by describing the items in a precise way.

The set of the first five odd numbers which is {1, 3, 5, 7, 9}, or the set of the names of the days of a week, are both clear and unambiguous whereas a set of colours or a set of people are not as yet clearly defined.

Having defined a set it should be clear whether an item belongs to it or not. The letter e clearly belongs to the set of vowels whereas the number 7 clearly does not belong to the set of even numbers. The symbols $\in$ and $\notin$ are used for *belongs to* and for *does not belong to*. Using these symbols the sentence:

Monday $\in$ {days of the week}

is read as

Monday belongs to the set of days of the week

whilst the sentence:

10 $\notin$ {2, 4, 6, 8}

is read as

10 does not belong to the set of the first four even numbers.

It is customary to use capital letters to denote a particular set and small letters to denote the members of the set. For example we could use the letter V to denote the set of *vowels* or the letter E to denote the set of the first four *even* numbers. Thus V = {a, e, i, o, u} and 10 $\notin$ E.

The universal set

The universal set is the set of all the items that we are considering at a particular moment of time. At one time we might wish to consider any positive number as a possible solution for a particular equation, whereas on another occasion we might only be considering whole numbers whether positive or negative. In the first case the universal set would be the set of positive numbers including fractions, whereas in the second case it would be the set of positive and negative whole numbers. The symbol for the universal set is U, or a curly ε which is called **ensemble**.

The empty set

An empty set is a set which contains no members.

The set of people living in your house who are over 6'7" tall could well be an empty set. In the same way the set of people living in your house under two years of age could also be an example of an empty set. The symbol for the empty set is { }, or ∅ which is sometimes pronounced *oink*.

Subsets

A subset is made up of some or all of the members of a set. The set consisting of 2 and 4 is a subset of the set of even numbers. The symbol used for *is a subset of* is ⊂ and for *is not a subset of* is ⊄.

Thus $\{2, 4\} \subset \{2, 4, 6, 8\}$, whereas $\{1, 2\} \not\subset \{2, 4, 6, 8\}$.

Consider the set P = {Ian, Mark, Paul}. One possible subset is 'the set of boys' names' which in this case is the whole set. Another possible subset is 'the set of girls' names' which in this case is the empty set.

There are eight possible subsets of P. These are:

{Ian, Mark, Paul}, {Ian Mark}, {Ian, Paul}, {Mark, Paul}, {Ian}, {Mark}, {Paul} and { }.

The complement of a set

The complement of a set consists of all those members of the universal set which do not belong to the set itself.

If the universal set is the set of whole numbers from 1 to 10 inclusive, and E is the set of even numbers, then the complement of E is {1, 3, 5, 7, 9}.

The symbol for complement is '. Thus $E' = \{1, 3, 5, 7, 9\}$.

The intersection of sets

The intersection of two sets is the set which consists of the common items which belong to both of the sets. The symbol for intersection is ∩.

Thus $\{1, 2, 3\} \cap \{1, 3, 5, 7, 9\} = \{1, 3\}$. 1 and 3 belong to both sets.

The union of sets

The union of two sets is the set which consists of all the items which belong to either or both of the two sets. the symbol for union is ∪.

Thus $\{1, 2, 3\} \cup \{1, 3, 5, 7, 9\} = \{1, 2, 3, 5, 7, 9\}$.

2 Solution sets

This is simply the name given to the set of solutions for a particular equation. It is a helpful idea since using set language allows us to talk about an empty set if there are no solutions in a particular situation.

Examples: If the universal set is the set of *whole* numbers, from 1 to 10 inclusive, find the solution sets for the equations:

(1) $2x + 3 = 17$ (2) $2x + 3 = 16$
(3) $2x + 3 > 17$ (4) $2x + 3 \leqslant 16$
(5) $x^2 - 5x + 4 = 0$ (6) $x^2 - 4x + 4 = 0$
(7) $x^2 - 4x + 5 = 0$ (8) $(x - 1)(x - 2)(x - 3) = 0$

(1) The solution set for $2x + 3 = 17$ is $\{7\}$
(2) The solution set for $2x + 3 = 16$ is $\{\ \}$ (no roots in universal set)
(3) The solution set for $2x + 3 > 17$ is $\{8, 9, 10\}$
(4) The solution set for $2x + 3 \leqslant 16$ is $\{1, 2, 3, 4, 5, 6\}$
(5) The solution set for $x^2 - 5x + 4 = 0$ is $\{1, 4\}$ (two real roots)
(6) The solution set for $x^2 - 4x + 4 = 0$ is $\{2\}$ (two equal roots)
(7) The solution set for $x^2 - 4x + 5 = 0$ is $\{\ \}$ (no real roots)
(8) The solution set for $(x - 1)(x - 2)(x - 3) = 0$ is $\{1, 2, 3\}$

Solution set for simultaneous equations
Solve the simultaneous equations: $x + y = 6$ and $y = 2x$.

The solution set for $x + y = 6$ contains the subset of ordered pairs:

$$\{(0, 6), (1, 5), (2, 4), (3, 3), (4, 2), (5, 1), (6, 0)\}$$

The solution set for $y = 2x$ contains the subset of ordered pairs:

$$\{(0, 0), (1, 2), (2, 4), (3, 6), (4, 8), (5, 10)\}$$

The solution set for the pair of simultaneous equations:

$$x + y = 6 \text{ and } y = 2x$$

is the point which is common to the two lines.

This point is the intersection of the two lines or the ordered pair which is common to the solution sets for the two separate equations.

$(2, 4)$ is the common ordered pair so $x = 2$, $y = 4$ is the required solution.

3   Sets and Venn diagrams

Often sets and subsets are shown on a Venn Diagram. This is a diagram where the universal set is usually shown by a rectangle, and each set is represented by a loop within that rectangle. Sets which have members in common are represented by overlapping loops.

Example 1: Represent the numbers from 1 to 10 and the set of even numbers on a Venn Diagram.

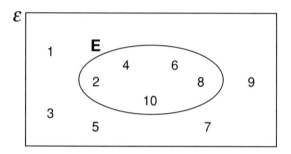

Note: the numbers outside the loop are the complement of the set of even numbers. So in this case $E' = \{1, 3, 5, 7, 9\}$.

Example 2: Represent the numbers from 1 to 20, the set of multiples of 3, and the set of multiples of 5 on a Venn Diagram.

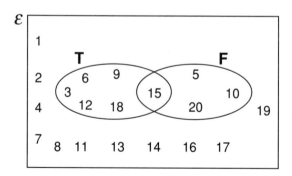

Note: the part where the two loops overlap is called the intersection of the two sets. In this case there is only one number, 15, in the intersection.

So in this case $T \cap F = \{15\}$

The numbers 5, 10 and 20 are the multiples of 5 which are not also multiples of 3. They are in the intersection of F and the complement of T.

This is written as F ∩ T′ = {5, 10, 20}.

Example 3: Represent the numbers from 1 to 40, the set of factors of 16, the set of factors of 24, and the set of factors of 36 on the same Venn Diagram.

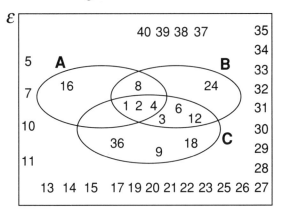

Note: the part which is common to all three sets contains those numbers which are factors of 16, 24 and 36. So in this case A ∩ B ∩C = {1, 2, 4}.

The numbers 3, 6 and 12 are factors of 24 and 36 but not of 16. This is shown using letters as B ∩ C ∩ A′ = {3, 6, 12}.

4 Set Algebra

The Venn diagrams which follow summarise the main ideas in set algebra together with a number of important properties.

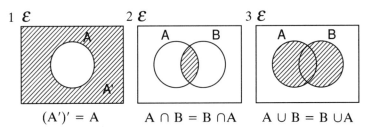

The results in 2 and 3 show the *commutative* property of ∩ and ∪.

4   $\mathcal{E}$

5   $\mathcal{E}$

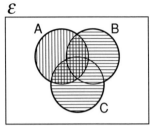

A ∩ (B ∩ C) = (A ∩ B) ∩ C   A ∪ (B ∪ C) = (A ∪ B) ∪ C

The results in 4 and 5 show the *associative* property of ∩ and ∪.

6   $\mathcal{E}$

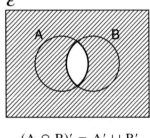

(A ∩ B)′ = A′ ∪ B′

7   $\mathcal{E}$

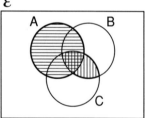

(A ∪ B)′ = A′ ∩ B′

The results shown in 6 and 7 are known as De Morgan's laws.

8   $\mathcal{E}$

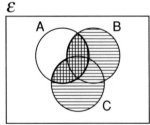

$\mathcal{E}$

A ∩ (B ∪ C) = (A ∩ B) ∪ (A ∩ C)
and A ∪ (B ∩ C) = (A ∪ B) ∩ (A ∪ C)

The results shown in 8 show that ∩ is *distributive* over ∪ and vice-versa.

Answers

Exercise 1 (p. 11)

1 (1) $100x$ (2) $\dfrac{n}{100}$

2 $\dfrac{100a}{n}$ **3** $100an + mb$

4 (1) $1000a$ (2) $3x$ (3) $\dfrac{y}{1000}$

5 $28 - n$ **6** $x + 50$, or $50 - x$ **7** $\dfrac{a}{x}$

8 (1) $x\,m$ (2) $\dfrac{180\ 000}{x}$

9 $2n + 5$ **10** (1) $x + 2, x - 2$ (2) $x + 1, x - 1$

11 $\dfrac{xa + yb}{100}$ **12** $\dfrac{2x + 5}{6y}$

13 $a - b$ **14** xy

15 $a + b$ **16** $\dfrac{x + 2}{y - 5}$

17 $mv + nu$ **18** $100a + b$

19 $xv; \dfrac{y}{v}$ **20** The original numbers are the digits of the final result, m and n.

Exercise 2 (p. 18)

1 (*a*) 10 doz.; 120   (*b*) $10a$; 120
2 (*a*) $5 \times 73 = 365$ (*b*) $5b = 365$
3 $3a + 4b$; 47   **4** $19a$; 47.5
5 (1) $26b$ (2) $19x$ (3) $12a$ (4) $4x$
6 (1) $2a + 4b$ (2) $12p + 3q$ (3) $6a + 3b + 4$

7 (1) $10a + 2b$ (2) $2b + c - d$ (3) $8x + 2z$
8 (1) 13   (2) 7   (3) 13   (4) 11
9 $8ab = 64$ (2) $3ax + bx = 42$
    (3) $3xy = 45$ (4) $4ab + 4bx - ay = 36$
10 $3\frac{3}{4}$ **11** 22
12 (1) 66   (2) 29   (3) 30
13 $3n + 12$ **14** $15a$
15 $a, a + d, a + 2d, a + 3d, a + 4d; 5a + 10d$
16 $2n + 5, 2n + 8, 2n + 11, 2n + 14, 2n + 17; 10n + 55$

Exercise 3 (p. 22)

1 $12a$ **2** $10xy$ **3** $2xy$

4 $21mn$ **5** $\frac{ab}{12}$ **6** $4ab$

7 $60abc$ **8** $\frac{xyz}{24}$ **9** x^3

10 a^4 **11** x^4 **12** a^6
13 $2a^5$ **14** $6x^7$ **15** $2a^2b^2$
16 $6b^4$ **17** x^3y^3 **18** $7x^6a^2$
19 $6a^6$ **20** $27a^6b^3$ **21** x^6
22 $8a^{12}$ **23** $16a^{12}$ **24** $64a^4$
25 54   **26** 21   **27** 8
28 172   **29** 33   **30** 1
31 216   **32** 576

Exercise 4 (p. 24)

1 a^2 **2** a^2 **3** $3x$ **4** $\frac{b^4}{2}$

5 $2a^4$ **6** $5y^3$ **7** $3x^3$ **8** $2c^2$
9 ab **10** x^3y **11** $5a^2b$ **12** $3xy$
13 2   **14** $\frac{5}{4}$

Exercise 5 (p. 27)

1 $\frac{12a}{35}$ **2** $\frac{x}{15}$

3 $\frac{4y + 5x}{xy}$ **4** $\frac{19a}{6b}$

5 $\dfrac{11x}{40y}$

6 $\dfrac{31x}{18y}$

7 $\dfrac{20a^3 + 9b^2c}{24a^2bc^2}$

8 $\dfrac{3x^2 + 4}{3x}$

9 $\dfrac{6ab^2 + 5}{2b}$

10 $\dfrac{40y^2 - 6}{5y^2}$

11 $\dfrac{a^2 + a + 1}{a}$

12 $\dfrac{x^2 - x + 1}{x}$

13 $\dfrac{c + 2a + 3b}{abc}$

14 $\dfrac{5bc - 4ac + 2ab}{a^2b^2c^2}$

15 x^2y

16 $\dfrac{4b}{a^2}$

17 $\dfrac{3b^2}{2a^2c}$

18 1

19 xy

20 $\dfrac{a}{bc}$

21 $\dfrac{4}{3a}$

22 $\dfrac{3y}{2}$

23 $\dfrac{9}{5q}$

24 $\dfrac{2xy}{3}$

25 $\dfrac{2x^2}{y}$

26 $2y$

27 1

28 $\dfrac{x^2}{y^2}$

29 $\frac{2}{3}$

30 $\dfrac{2x^4}{3y^2}$

31 $\dfrac{x^2}{y^2}$

32 $\dfrac{2a^3}{3b^2}$

33 $\dfrac{8a^2}{3c}$

34 $\dfrac{3y^2}{4}$

35 $\dfrac{20}{xy}$

36 $\dfrac{4a^2b^2}{c^4}$

Exercise 6 (p. 33)

1 $15x + 18z$

2 $6a^2 + 8ab$

3 $18a^3 + 42a^2b - 36a^2c$

4 $8x + y$

5 $5x^3 + 4x^2$

6 x

7 $8x + 5y - 4z$

8 $x - 2y - z$

9 $2x - y + 2z$

10 $a + 7b$

11 $a - b$

12 $a + b$

13 $4x + 2y - 2z$

14 $a + 5b - 5c$

15 $3y$

16 $a^2 + b^2$

17 $x^3 + x^2y - x^3y + xy^3$

18 $x^2 + 9x + 23$

19 $9p^2 + 3pq + 15q^2$

20 $5x^2y^2 - 3xy + 6x^2$

21 $2x^4 + 8x^3$

Exercise 7 (p. 35)

1 $6a - 9$ **2** $3a + 30b$ **3** $\frac{3}{2}$

4 $3a^2 + 2ab - 2ac$ **5** $2p^2 + p$

6 $x^2 + 3xy + 12y^2$ **7** $7bc - 2b^2 + 2c^2$

8 $13x + 5$ **9** $34 - 6a + 4b$ **10** $x + y$

11 $2a - 10b$ **12** $\frac{3x^2}{2y}$ **13** $\frac{27c}{8}$

14 $2a - (b - c)$ **15** $x - (y + z)$ **16** $2(a + 2b - 3c)$

17 $x^2 - y(x - y)$ **18** $2a - 4b + 7c$ **19** $x + 2y - 2x$

20 (1) 111   (2) 6

Exercise 8 (p. 43)

1 $0 + 4 - 2 + 4 - 6 = 0$

2 $+8 + 2 - 14 + 4 - 6 = -6°C$

3 (1) 11°C   (2) 15°C   (3) 30°C

4 (1) $+5$ (2) -5 (3) $+8$

5 (1) (*a*) $+2$ (*b*) $+4$ (2) $+5$

6 (1) $(+4)$ (2) $(+8)$ (3) (-8)

    (4) (-4) (5) $(+3)$ (6) (-8)

    (7) 0   (8) (-8)

7 (1) $7a$ (2) $-7x$ (3) $10ab$

    (4) $-x - 8y$ (5) $a - 7b$ (6) $-x + 2y$

    (7) $7x - 3y$ (8) -8

8 (1) $2x - 2y$ (2) $2x - y - 7z$

9 (1) $3a - (-5a)$ (2) $5x - (6x)$

    (3) $-3a - (-10a)$

10 (1) $(-a)$ (2) $(+a)$ (3) $(-a)$

Exercise 9 (p. 47)

1 (1) $(+36)$ (2) (-36) (3) (-36)

    (4) $(+36)$ (5) $(+4)$ (6) (-4)

    (7) (-4) (8) $(+4)$

2 (1) $(-a^2)$ (2) $(+a^2)$ (3) (-1)

    (4) $(+1)$

3 (1) $(-4ab)$ (2) $(+4ab)$ (3) $(-20xy)$

    (4) $(-20xy)$ (5) $\left(-\dfrac{5x}{y}\right)$ (6) $\left(+\dfrac{5x}{y}\right)$

4 (1) $+24$ (2) $+6a^3$

    (3) $+3y$ (4) $-6ab$

5 (1) $-10x^3$ (2) $a^2 - b^2$

(3) $-2ab^2$ (4) $-a^2 + 2ab + ac$

6 (1) $+20ab^2$ (2) $6x^2 + 8x$

(3) $-2xz + 2yz$

7 (1) $a^2, -a^3, a^4, -a^5$

(2) $4x^2, -8x^3, +16x^4, -32x^5$

(3) $\dfrac{b^2}{9}, -\dfrac{b^3}{27}, \dfrac{b^4}{81}, \dfrac{b^5}{243}$

8 $\pm 9, \pm 3x^2, -x, -2a^2$

9 (1) $+16x$ (2) $5x$ (3) $2y$

(4) $-2b$ (5) $-2t$ (6) $2x$

(7) $8x^3$ (8) $-3x$ (9) $-4ab$

(10) $6a^2c$

10 (1) $-b$ (2) $-x$ (3) $-\dfrac{9x}{4}$

Exercise 10 (p. 51)

1 (*a*) 17   (*b*) 47   (*c*) 20   (*d*) 47

2 (*a*) 3n   (*b*) 3n − 1   (*c*) 3n − 2   (*d*) n^2

(*e*) $n^2 + 10$ (*f*) 5n + 1

Exercise 11 (p. 52)

1 (*a*) 1, 5, 9, 13, 17   (*b*) 16, 20, 24, 28, 32

(*c*) 4, 16, 36, 64, 100

2 (*a*) −3, 1, 5, 9, 13   (*b*) 32, 52, 72, 92, 112

(*c*) 32, 44, 64, 92, 128

3 (*a*) 18, 23, 28, 33, 38   (*b*) 11, 23, 35, 47, 59

(*c*) 19, 34, 55, 82, 115

4 (*a*) 17, 24, 31, 38, 45

(*b*) 17, 24, 31, 38, 45; yes; 7(n + 1) + 3 = 7n + 7 + 3

$$= 7n + 10$$

Exercise 12 (p. 53)

1 (*a*) 5(4n − 3)   (*b*) 5(n + 3) − 2   (*c*) 2(6n − 5) − 7

(*d*) 2(3(n + 4) − 5)

2 (*a*) 11   (*b*) 59   (*c*) 93

3 (*a*) 17   (*b*) 17

Exercise 13 (p. 56)

1 (a) $x \to 2x + 7$; 13 (b) $x \to 2(x + 7)$; 20
  (c) $x \to (x + 5)^2$; 64 (d) $x \to x^2 + 5$; 14
2 (a) 2 (b) 18
  (c) $x \to 2x - 3$ (d) $x \to 2(x - 3)$
3 (a) 40 (b) 47 (c) 361 (d) 4101
  (e) $x \to (2x + 7)^2 + 5$

Exercise 14 (p. 57)

1 (a) 23 (b) $x \to (x - 3) \div 4$
2 (a) 23 (b) $x \to (x - 3) \div 5 + 4$
3 $x \to (x + 3) \div 2$; 7
4 $x \to ((x - 7) \div 5 + 3) \div 2$; 3

Exercise 15 (p. 59)

1 (a) 4 (b) 10 (c) 25 (d)
2 (a) 6 (b) 5
3 (a) 5 (b) 3

Exercise 16 (p. 65)

1 (a) $2\frac{1}{2}$ (b) 20
2 (a) 16 (b) 60
3 (a) -16 (b) -10
4 (a) 80 (b) -72
5 (a) 12.6 (b) -0.08
6 (a) $3\frac{1}{5}$ (b) 5
7 (a) 2.5 (b) 12
8 (a) 8 (b) $\frac{4}{3}$
9 (a) $\frac{8}{5}$ (b) 16
10 (a) 1.8 (b) -45
11 (a) 14.8 (b) $18\frac{3}{4}$
12 (a) $\frac{10}{9}$ (b) $\frac{1}{3}$
13 22 **14** 5 **15** 8 **16** 29
17 $-\frac{1}{2}$ **18** -3 **19** $18\frac{4}{5}$
20 (a) $\frac{32}{5}$ (b) $27\frac{1}{2}$
21 (a) -80 (b) -10
22 -35 **23** 8 **24** -7 **25** $\frac{7}{3}$
26 3 **27** $\frac{7}{3}$ **28** $-\frac{55}{2}$ **29** 2.7
30 2.9 **31** 3 **32** $\frac{45}{19}$ **33** 36
34 $16\frac{1}{2}$

Exercise 17 (p. 68)

1 12	**2** 72.5		**3** 42
4 15	**5** 13.5		**6** 21, 23, 25
7 10 km		**8** 40	
9 11.9 cm, 10.1 cm		**10** 4 men, 60p	
11 5		**12** Son 8 years, father 32 years	
13 30		**14** 80 at 18p and 40 at 12p	
15 23			

Exercise 18 (p. 71)

1 175	**2** 154		**3** $113\frac{1}{7}$
4 85.6	**5** 0.616		**6** $57\frac{5}{7}$
7 $4\frac{1}{2}$	**8** 204		**9** $\frac{5}{8}$
10 1390 (approx.)	**11** 31.4		

Exercise 19 (p. 75)

1 $r = \sqrt{\dfrac{A}{\pi}}$

2 $r = \sqrt[3]{\dfrac{3V}{4\pi}}$

3 $r = \sqrt{\dfrac{3V}{\pi h}}$

4 $C = \dfrac{825H}{E}$

5 $B = \sqrt{\dfrac{F \times 112 \times 10^5}{A}}$

6 $l = \sqrt[3]{\dfrac{48EId}{W}}$

7 $s = \dfrac{v^2 - u^2}{2f}$, $17\frac{1}{2}$

8 (1) $V = IR$ (2) $R = \dfrac{V}{I}$, 0.1

9 $n = \sqrt{\dfrac{NR - 1}{r}}$, $\frac{70}{9}$

10 $k = 1008$; $v = \dfrac{1008}{p}$

Exercise 20 (p. 77)

1 $\dfrac{4a}{5}$

2 $2a$

3 p

4 $4b$

5 $\dfrac{3a - 2b}{a}$

6 $\dfrac{bp + c}{b}$

7 $\dfrac{2(b - a)}{b}$

8 $\dfrac{6 - 19b}{3a}$

9 p

10 $\dfrac{2b - 12}{3a}$

11 $\dfrac{3b}{a - b}$

12 $\dfrac{12a}{4 + 3a}$

13 $\dfrac{a^2 + b^2}{a - b}$

14 $\dfrac{m}{n - 3}$

Exercise 21 (p. 84)

1	$x = 3, y = 6$	**2**	$x = 5, y = 8$
3	$x = 12, y = 3$	**4**	$x = 8, y = 3$
5	$x = 3, y = 1$	**6**	$x = 4, y = 5$
7	$x = 7, y = 4$	**8**	$x = 10, y = 7$
9	$x = 6, y = 5$	**10**	$x = 3, y = 10$
11	$x = 6, y = 10$	**12**	$a = 5, b = -\frac{3}{2}$
13	$a = -2, b = 3$	**14**	$p = -2, q = 4$
15	$x = \frac{5}{2}, y = -2$	**16**	$x = 9, y = 13$
17	$a = -7, b = \frac{3}{2}$	**18**	$x = 6, y = -4$
19	$x = 8, y = 12$	**20**	$x = 1.35, y = 2.7$
21	$x = 4, y = 2$	**22**	$P = 1.8, Q = 0.32$
23	$x = \frac{15}{2}, y = -3$	**24**	$x = 12, y = 6$

Exercise 22 (p. 87)

1 $x = 12, y = 10$ **2** $16, 12$

3 $20, 7$ **4** $10, 6$

5 $m = 2, b = -3$. Equation is $y = 2x - 3$; when $x = 6$. $y = 9$.

6 $m = 20, b = 10$

7 $a = 0.5, b = 0.6$ **8** $a = -1.36, b = 1.38$

9 10 m, 6 m **10** £3.75; 2p

11 Tie 49p, socks 24p **12** $u = 10, a = 6, 125$ m

Exercise 23 (p. 92)

1 (*a*) $H < 190$ cm (*b*) $P > 50$

(*c*) £200 $\leq S <$ £300 (*d*) 3 hours $< T <$ 5 hours

Exercise 24 (p. 95)

1 (*a*) $x > 7$ (*b*) $x < 3$

2 (*a*) $x < 8$ (*b*) $x \leq 3$

3 (*a*) $x < 8$ (*b*) $x \leq 5$

4 (*a*) $x < 5$ (*b*) $x \geq 2$

5 (*a*) $x < 3$ (*b*) $x < -15$

6 $x > 4$

7 $x \geq 8$

8 $x < 16$

Exercise 25 (p. 97)

1 (*a*)]2, 7[(*b*) [2, 7]   (*c*) [2, 7[(*d*)]2, 7]
2 (*a*) 3 < x < 6   (*b*) 5 ⩽ x < 8
    (*c*) −2 < x ⩽ 5   (*d*) −2 ⩽ x < 5
3 0 ⩽ x ⩽ 5
4 2 ⩽ x < 5
5 3 < x < 7

Exercise 26 (p. 98)

1 (*a*) 4 < x < 7   (*b*) −2 < x < 1   (*c*) 4 < x ⩽ 8
2 −3 ⩽ x < 3

Exercise 27 (p. 107)

2 10.4°C   **3** Males: (1) 33.9   (2) 18.1
                  Females: (1) 36.8   (2) 20.2
4 (1) 44.2 kg   (2) 65 kg   (3) 1.5 kg
5 Accurate deductions simply on the basis of the graphs (which are anyway not regular) are certain to be unrealistic. However . . .
    (1) 1978–9   (2) 126 000   (3) 1978–9
6 25.1 m/s
7 P = £13.30 is wrong and should be £15. £12.80; £2.50; £36.50
8 About 360 kilotherms.
9

x	−3	−2	−1	0	1	2	3	4	5
y	6	2.5	0	−1.5	−2	−1.5	0	2.5	6

3.45; −1.45.

Exercise 28 (p. 116)

1 (1) 2   (2) −2   **2** (1) −1   (2) $\frac{2}{3}$
3 (1) $-\frac{5}{4}$ (2) $\frac{5}{6}$ **4** (1) 3   (2) 2
5 (1) 4   (2) 6   **6** (1) $-\frac{1}{2}$ (2) 1
7 a = 2, b = 3. Equation is y = 2x + 3. Intercept is 3.
8 x + 2y = 4

Exercise 29 (p. 119)

1 A, (4, 4);   B, (4.6, 1.2);   C, (−2, 3);
   D, (4, −2);   E, (−1.4, −3.4);   F, (3, 0);
   G, (0, −3).
2 (1.12, 1.12)
4 (2, 2)
5 They lie on a straight line parallel to OX.
6 They appear to lie on a straight line which passes through the origin.
7 For every point on it the x co-ordinate is +3.

Exercise 30 (p. 126)

1 All lines pass through the origin with different slopes.
2 All lines are parallel with slope 45°.
3 All lines are parallel with gradient $\frac{1}{2}$.
4 All lines have the same intercept on the y axis, viz. 2.
5 Intercepts are: (1) 1.5 and 3   (2) 4 and −2
   (3) 2 and 5   (4) 2.5 and −2
6 (1) x = 4, y = 3   (2) x = 3, y = 1
   (3) x = −2, y = 2
7 a = 3; −1
8 b = 1; 1
9 a = 3, b = −2. Equation is y = 3x − 2; −2.
10 y = 2x + 3
11 3y − 2x = 7. Gradient = $\frac{2}{3}$.
12 y − x = 2. Gradient = 1.

Exercise 31 (p. 131)

1 (*a*) yes   (*b*) no   (*c*) yes   (*d*) no
2 (2, 5) is in the region; (6, 1) is not in the region.
3 (3, 4) is in the region; (4, 6) is not in the region.
5 (3, 6) is not in the region; (3, 7) is in the region.
6 (*a*) x + y ⩽ 36, y ⩾ 2x + 3, x + 3y ⩽ 12
   (*b*) x + 3y ⩽ 12, y ⩽ 2x + 3, y + y ⩽ 6
   (*c*) x + 3y ⩽ 12, x + y ⩾ 6

Exercise 32 (p. 133)

1 (5, 7)
2 (3, 4)

3 (3, 5)
4 (6, 3) and (5, 4); (6, 4) is not included since x + 3y<18.
5 (8, 4); (7, 4) is not included since y<2x − 10

Exercise 33 (p. 136)

1 (*a*) 8L + 3S $\geqslant$ 144   (*b*) L + S $\leqslant$ 36; £(32L + 8S)
2 L + S $\leqslant$ 100; 18L + 45S $\leqslant$ 2700
3 8000A + 9600B $\leqslant$ 72 000; 48A + 30B $\leqslant$ 360; A $\geqslant$ 4; B $\geqslant$ 2
4 (*a*) V + L $\leqslant$ 10   (*b*) 36V + 90L $\leqslant$ 540
    (*c*) 600V + 1100L $\geqslant$ 6600
5 (16, 2), (14, 3) and (12, 4)

Exercise 34 (p. 143)

1 $ab + ay + bx + xy$
2 $ce + cf + de + df$
3 $acxy + adx + bcy + bd$
4 $ab - ay - bx + xy$
5 $ax - bx - ay + by$
6 $ab + ay - bx - xy$
7 $ab - ay + bx - xy$
8 $ab + 3a + 2b + 6$
9 $ab - 3a - 2b + 6$
10 $ab + 3a - 2b - 6$
11 $ab - 3a + 2b - 6$
12 $x^2 + 12x + 35$
13 $a^2b^2 + 9ab + 18$
14 $x^2 + 13x + 30$
15 $x^2 - 13x + 30$
16 $x^2 + 7x - 30$
17 $x^2 - 7x - 30$
18 $p^2 - 4p - 96$
19 $x^2 - 12xy + 32y^2$
20 $x^2 + 4xy - 32y^2$
21 $x^2 - 4xy - 32y^2$
22 $2a^2 + 9ab + 10b^2$
23 $9x^2 - 27xy + 20y^2$
24 $28x^2 + 15x + 2$
25 $6x^2 - 11x + 3$
26 $9x^2 - 9x - 4$
27 $1 - y - 12y^2$
28 $18x^2 - 27x - 5$
29 $14x^2 + 29xy - 15y^2$
30 $18a^2 - 57ab + 35b^2$
31 $a + b$
32 $a + b - c$
33 12
34 $5y - 5x = 0.5$
35 (*a*) $x^3 - y^3$
    (*c*) $1 + x^3$
    (*b*) $a^3 + 8$
    (*d*) $x^3 + 3x^2a + 3xa^2 + a^3$

Exercise 35 (p. 146)

1 $x^2 + 4x + 4$
2 $x^2 - 4x + 4$
3 $a^2 + 6ab + 9b^2$
4 $a^2 - 6ab + 9b^2$
5 $4x^2 + 4xy + y^2$
6 $x^2 - 4xy + 4y^2$
7 $a^2b^2 + 20ab + 100$
8 $x^2y^2 - 6xy + 9$
9 $16x^2 + 40xy + 25y^2$
10 $16x^2 - 40xy + 25y^2$
11 $25x^2y^2 + 60xy + 36$
12 $1 - 20x^2 + 100x^4$

13 $25x^4 + 30x^2y^2 + 9y^4$ **14** $9x^2y^2 - 12xy^3 + 4y^4$

15 $x^2 + \dfrac{2x}{y} + \dfrac{1}{y^2}$ **16** $\dfrac{1}{x^2} - \dfrac{2}{xy} + \dfrac{1}{y^2}$

17 $a^2 + \dfrac{2a}{3} + \frac{1}{9}$ **18** $\frac{1}{4} - \dfrac{3y}{4^2} + \dfrac{9y^2}{16}$

19 $x^2 + 2xy + y^2 + 2x + 2y + 1$
20 $1 - 2x + 4y + x^2 - 4xy + 4y^2$
21 $a^2 + b^2 + c^2 + 2ab - 2ac - 2bc$
22 $x^2 + y^2 + z^2 - 2xy + 2xz - 2yz$
23 $4x^2 + 9y^2 + 25^2 + 12xy - 20xz - 30yz$
24 $16a^2 + 4b^2 + 1 - 16ab - 8a + 4b$
25 $x^3 + 3x^2y + 3xy^2 + y^3$ **26** $x^3 - 3x^2y + 3xy^2 - y^3$
27 $a^3 + 6a^2 + 12a + 8$ **28** $a^3 - 6a^2 + 12a - 8$
29 $p^3 + 3p^2q + 3pq^2 + q^3$ **30** $p^3 - 3p^2q + 3pq^2 - q^3$
31 $8x^3 + 12x^2y + 6xy^2 + y^3$ **32** $x^3 - 6x^2y + 12xy^2 - 8y^3$
33 $27a^3 - 27a^2 + 9a - 1$ **34** $1 - 9b + 27b^2 - 27b^3$
35 $2xy$ **36** (1) $4ab$; (2) $-4ab$
37 $40x$ **38** $9y^2 + 18y + 9$; 36
39 $4a(x + a)$; 256 m^2 **40** $2a(x + y + 2a)$

Exercise 36 (p. 147)

1 $a^2 - x^2$ **2** $p^2 - q^2$
3 $a^2 - 4b^2$ **4** $16x^2 - 9$
5 $4x^2 - 1$ **6** $1 - 36x^2$
7 $1 - a^4$ **8** $4x^4 - 1$
9 $x^4 - y^4$ **10** $9x^2y^2 - 4$
11 $144x^2y^2 - 1$ **12** $\frac{1}{16}x^2 - 49$
13 $(x + y)^2 - z^2$ **14** $(a + x)^2 - y^2$
15 $(2a + 3b)^2 - 1$ **16** $(x - 2y)^2 - 36$
17 $a^2 - 4(b + c)^2$ **18** $4x^2 - 9(y + z)^2$
19 $x^2 - \frac{4}{9}$ **20** $\dfrac{x^2}{4} - \dfrac{y^2}{9}$
21 $(x + a)^2 - \frac{1}{4}$

Exercise 37 (p. 149)

1 $6(x + 2)$ **2** $a(3b + 2)$
3 $2y(2x + y)$ **4** $2a(3a - 2b)$
5 $7xy(2xy - 1)$ **6** $16(1 - 2a^2)$
7 $a(a - b + c)$ **8** $x(x^2 + 3x - 1)$
9 $a^2(15a - 5b + 3b^2)$ **10** $3ac(2a - 5c)$

11 $ab(a + b - c)$

12 $\dfrac{bc}{3}\left(\dfrac{c}{2} - \dfrac{a}{3}\right)$

13 $7.4(13^2 + a^2)$

14 $18.6(18.6 + 1.4) = 18.6 \times 20 = 372$

Exercise 38 (p. 152)

1 $(a + b)(x + y)$
2 $(p + q)(c + d)$
3 $(a + d)(b + e)$
4 $(x - y)(a - c)$
5 $(x + p)(x + q)$
6 $(x - g)(x - h)$
7 $(a + 5)(b + 6)$
8 $(a - 5)(b - 6)$
9 $(a - 5)(b + 6)$
10 $(2a + 3)(b - 5)$
11 $(ax - b)(x + a)$
12 $(x - b)(x + a)$

Exercise 39 (p. 154)

1 $(x + 2)(x + 1)$
2 $(x - 2)(x - 1)$
3 $(x + 3)(x + 2)$
4 $(x - 3)(x - 2)$
5 $(x + 6)(x + 1)$
6 $(x + 5)(x + 4)$
7 $(x - 10y)(x - 2y)$
8 $(a - 12b)(a - 3b)$
9 $(xy + 6)(xy + 9)$
10 $(ay - 16)(ab - 3)$
11 $(y - 9)(y - 12)$
12 $(x - 7y)(x - 5y)$
13 $(x - 2)(x + 1)$
14 $(x + 2)(x - 1)$
15 $(x + 3y)(x - 2y)$
16 $(x - 3y)(x + 2y)$
17 $(b - 3)(b + 1)$
18 $(b + 3)(b - 1)$
19 $(x + 16)(x - 3)$
20 $(x - 16)(x + 3)$
21 $(x - 11y)(x + 10y)$
22 $(a - 12)(a + 1)$
23 $(a - 4)(a + 3)$
24 $(p + 9)(p - 8)$
25 $(p - 36)(p + 2)$
26 $(1 - 5x)(1 - 4x)$
27 $(1 - 10x)(1 + 2x)$
28 $(xy - 11)(xy + 8)$
29 $(p + 9)(p - 5)$
30 $(p - 7q)(p + 8q)$

Exercise 40 (p. 155)

1 $(3x + 4)(x + 2)$
2 $(4x - 3)(3x - 2)$
3 $(6x + 1)(2x - 5)$
4 $(9x - 2)(x + 5)$
5 $(2x + 1)(x + 1)$
6 $(3x - 1)(x - 1)$
7 $(2x + 1)(x + 2)$
8 $(3x + 1)(2x + 1)$
9 $(2x - 3)(2x - 1)$
10 $(5x - 1)(x - 1)$
11 $(2x - 3)(3x - 1)$
12 $(4x + 1)(3x + 2)$
13 $(2a - 1)(a + 1)$
14 $(2a + 1)(a - 1)$
15 $(2a + 3)(a - 2)$
16 $(5b + 2)(2b - 1)$

17 $(5b - 2)(2b + 1)$ **18** $(2y - 5)(4y + 3)$
19 $(4x - 1)(3x + 2)$ **20** $(7c + 2)(2c - 3)$

Exercise 41 (p. 158)

1 $(p + q)^2$ **2** $(x - 2y)^2$
3 $(3x + 1)^2$ **4** $(4x - 5y)^2$

5 $(x + \frac{1}{2})^2$ **6** $\left(\dfrac{a}{3} + \dfrac{b}{2}\right)^2$

7 $(a + b + 2)^2$ **8** $(x - y - 5)^2$
9 $(x + 10)(x - 10)$ **10** $(ab + 5)(ab - 5)$
11 $(2x + 3y)(2x - 3y)$ **12** $(5a + 4b)(5a - 4b)$
13 $(11x + 6y)(11x - 6y)$ **14** $(12p + 13q)(12p - 13q)$
15 $(5 + 4a)(5 - 4a)$ **16** $(1 + 15x)(1 - 15x)$
17 $2(2a + 5b)(2a - 5b)$ **18** $3(x + 5)(x - 5)$
19 $5(x + 3y)(x - 3y)$ **20** $(a + b + c)(a + b - c)$
21 $(x + 2y + 4z)(x + 2y - 4z)$
22 $(1 + \frac{9}{4}y)(1 - \frac{9}{4}y)$ **23** $(x + y + z)(x - y - z)$
24 $(a + x - 2y)(a - x + 2y)$ **25** $(x - 15)(x - 1)$
26 $4ab$ **27** 3000
28 2100 **29** 880
30 252 **31** 15
32 150 **33** 264
34 68 **35** 38.4
36 630 **37** 140
38 (1) 20, (2) 60

Exercise 42 (p. 160)

1 $(x + c)(x^2 - cx + c^2)$ **2** $(y - a)(y^2 + ay + a^2)$
3 $(1 + 2a)(1 - 2a + 4a^2)$ **4** $(x - 4)(x^2 + 4x + 16)$
5 $(2 + 3c)(4 - 6c + 9c^2)$ **6** $(R - 1)(R^2 + R + 1)$
7 $(m - 5n)(m^2 + 5mn + 25n^2)$

8 $(xy + \frac{1}{2})\left(x^2y^2 - \dfrac{xy}{2} + \dfrac{1}{4}\right)$

9 $\left(\dfrac{1}{x} + \dfrac{1}{y}\right)\left(\dfrac{1}{x^2} - \dfrac{1}{xy} + \dfrac{1}{y^2}\right)$

10 $\left(\dfrac{1}{x} - \dfrac{1}{y}\right)\left(\dfrac{1}{x^2} + \dfrac{1}{xy} + \dfrac{1}{y^2}\right)$

Exercise 43 (p. 163)

1 $\dfrac{y}{3xz}$

2 $\dfrac{3d^2}{4a^2}$

3 $\dfrac{x(x + y)}{2y(x - y)}$

4 $\dfrac{a^2 + 3b}{2ab(a - b)}$

5 $\dfrac{3}{2xy}$

6 $\dfrac{a + b}{a}$

7 $\dfrac{2}{x - 2}$

8 $\dfrac{x - 5}{x + 5}$

9 $\dfrac{a(a + 3b)}{b(a + 2b)}$

10 $\dfrac{x + y}{x^2 + xy + y^2}$

Exercise 44 (p. 164)

1 $\dfrac{x^2}{(x + 1)}$

2 $y(y + 1)$

3 $\dfrac{b}{a}$

4 $\dfrac{a - 7}{a - 3}$

5 $\dfrac{a(a + 2b)}{a - b}$

6 $\dfrac{(x + 1)(x - 3)}{x}$

7 1

8 $\dfrac{1}{2(x - 1)}$

9 $\dfrac{2(b - 4)}{b(b - 1)}$; $\frac{1}{10}$

10 $\dfrac{x + 3}{2}$; 2.25

Exercise 45 (p. 166)

1 $\dfrac{2x - y}{x(x - y)}$

2 $\dfrac{-(x + 1)}{(x - 1)(x - 2)}$

3 $\dfrac{13a - 5b}{6(a^2 - b^2)}$

4 $\dfrac{x^2 + 2x - 2}{x(x + 2)}$

5 $\dfrac{3x^2 - 20x + 34}{(x - 3)^2}$

6 $\dfrac{-5x^2 - 17x}{6(x + 1)(x - 2)}$

7 $\dfrac{7 - 3a}{(1 - a)^2}$

8 $\dfrac{12x}{x^2 - 9}$

9 $\dfrac{2y - x^2 - 2xy}{x(x + 2y)}$

10 $\dfrac{a(10b - 3a)}{3b(3a - b)}$

11 $\dfrac{t(a - b)}{(1 - at)(1 - bt)}$

12 $\dfrac{(x - a)(\cancel{\quad})t(y - b)}{(a + bt)(x + yt)}$

13 $\dfrac{2x - 1}{x^2 - y^2}$

14 $\dfrac{-2y^2}{(x - y)^2(x + y)}$

15 $R = \dfrac{pq}{p + q}$

16 $R = \dfrac{pq}{q - p}$

17 $R = \dfrac{R_1R_2}{2R_2 + 3R_1}$

18 $R = \dfrac{R_1R_2}{3R_2 - 2R_1}$

19 $R = -\dfrac{r^2 - s^2}{2s}$

20 $R = \dfrac{5(p^2 - q^2)}{5p - q}$

21 $\dfrac{2PQ}{P + Q}$

22 $\dfrac{Q(P + Q)}{P - Q}$

23 1.47

24 $P = \dfrac{2b + a}{2a + b} \cdot Q$

25 $n = \dfrac{IR}{E - Ir}$; 6

Exercise 46 (p. 168)

1	3	**2**	$\frac{5}{6}$	**3**	2.5
4	$-\frac{3}{5}$	**5**	$8\frac{3}{19}$	**6**	$\frac{5}{3}$
7	7	**8**	0.3	**9**	4
10	6	**11**	$\frac{8}{5}$	**12**	6
13	8	**14**	$V = \dfrac{RE}{R + r}$		

Exercise 47 (p. 183)

Answers involving decimals are mostly approximate.

1 (1) 5.3   (2) $\pm$ 2.65   (3) $\pm$ 1.87

2 (1) $\pm$ 1.8   (2) $\pm$ 2.83   (3) $\pm$ 3.46

3 (1) $\pm$ 2.45   (2) $\pm$ 2.24   (3) $\pm$ 1.31

4 (1) $\pm$ 1.41   (2) $\pm$ 2.45

5 (1) 3.73 or 0.27   (2) -4 or -2
   (3) 4.24 or -0.24

6 (1) -3.41 or -0.59 (2) -3 or -1
   (3) 0.83 or -4.83

7 Min. value is -4 when x = 3.
   (1) 5 or 1   (2) 6.16 or -0.16

8 Min. value is -2 when x = 2.
   (1) 4.45 or -0.45 (2) 5 or -1

9 Min. value is -1.1 approx. when $x = \frac{5}{4}$. On substituting $x = \frac{5}{4}$ in the function, this equals $-1\frac{1}{8}$.
   (1) 2 or $\frac{1}{2}$ (2) 2.69 or -0.19

10 Max. value is 2.25 when x = -0.5.

Exercise 48 (p. 185)

1 (a) $x < 1$ or $x > 2$ (b) $1 < x < 2$
2 (a) $x \leq 3$ or $x > 1$ (b) $-3 < x < 1$
3 $x < -1$ or $x > 4$
4 $-5 \leq x \leq 1$

Exercise 49 (p. 186)

1 $5 > 1 - 4 + 3 \, (= 0)$
2 $5 > 1 - 4 - 5 \, (= -8)$
3 $3 > 1$ and $3 < 5$
4 $(3, 1)$

Exercise 50 (p. 191)

1 ± 2 **2** $\pm \frac{1}{2}$
3 ± 12 **4** $1, -3$
5 $8, -2$ **6** $1, -11$
7 $\frac{3}{4}, -\frac{5}{4}$ **8** $16.75, 1.25$
9 $8, 2$ **10** $3, -4$
11 $5, -3$ **12** $4, -7$
13 $8, -4$ **14** $2, \frac{3}{2}$
15 $\frac{5}{2}, -1$ **16** $3.25, -0.92$
17 $1.434, 0.232$ **18** 4 and -1.5
19 $5, -\frac{1}{3}$ **20** $2.732, -0.732$
21 $13.14, -1.14$ **22** $6.59, -7.59$

Exercise 51 (p. 193)

1 $0, 3$ **2** $0, -5$ **3** $2, 2$
4 $1, 2$ **5** $1, -4$ **6** $3, -7$
7 $\frac{7}{3}, -\frac{11}{2}$ **8** $4, 5$ **9** $2, -3$
10 $5, -7$ **11** $-3, -10$ **12** $11, -6$
13 $9, -5$ **14** $\frac{3}{2}, 4$ **15** $\frac{5}{2}, -1$
16 $1, \frac{1}{3}$ **17** $\frac{5}{2}, -\frac{3}{4}$ **18** $\frac{1}{4}, -\frac{2}{3}$
19 $2, 4, 5$ **20** $\frac{1}{2}, -3, -2$ **21** $1, 4, -2$
22 $\frac{5}{2}, 10, -5$ **23** a, b **24** $\frac{c}{2}, -d$

Exercise 52 (p. 197)

1	$0.303, -3.303$	**2**	$4.561, 0.438$
3	$1.48, -1.08$	**4**	$1.43, 0.23$
5	$2.85, -0.35$	**6**	$1.744, -0.344$
7	$0.535, -6.535$	**8**	$3.441, -0.775$
9	$0.427, -2.927$	**10**	$0.803, -2.803$
11	$1.081, -1.481$	**12**	$-0.130, -0.770$
13	$2.591, -0.257$	**14**	$3.68, -0.43$
15	$9.75, -1.75$	**16**	$2.13, -5.62$
17	$6.123, -2.123$		

Exercise 53 (p. 199)

1 $\frac{5}{2}$ or $\frac{2}{5}$ **2** 9 mm, 15 mm
3 6.53 or -1.53 **4** 21.37, 27.37
5 5.55. Negative root has no meaning for the problem.
6 79.78. Second root is inadmissible.
7 $12.5, -10$
8 8. Negative root is inadmissible.
9 15 mm, 20 mm   **10** 144 m^2, 225 m^2
11 $7, -1$
12 12. Negative root is inadmissible.
13 108 m^2 **14** 78.4 m
15 8 and 12, or -12 and -8 **16** 12

Exercise 54 (p. 202)

1 $x = 6, y = 4; x = -5, y = -7$
2 $x = 5, y = 2; x = -26, y = 33$
3 $x = 2, y = 1; x = -\frac{7}{11}, y = \frac{69}{11}$
4 $x = \frac{1}{2}, y = 6\frac{1}{2}; x = 3\frac{1}{2}, y = -2\frac{1}{2}$
5 $x = -3, y = 2; x = -2, y = 1$
6 $x = 1, y = 4; x = \frac{4}{3}, y = \frac{34}{9}$
7 $x = 4, y = 3; x = -\frac{15}{14}, y = -\frac{171}{14}$
8 $x = -14, y = 11; x = \frac{9}{2}, y = -\frac{4}{3}$
9 $x = 7, y = 4; x = \frac{4}{3}, y = 21$
10 $x = 6, y = 4; x = -\frac{6}{13}, y = -\frac{116}{13}$
11 $x = 2, y = 4; x = \frac{4}{9}, y = -\frac{2}{3}$

Exercise 55 (p. 204)

1 $x = 3, y = 5; x = 5, y = 3$

2 x = 7, y = 2; x = 2, y = 7
3 x = 8, y = 3; x = −3, y = −8
4 x = ± 4, y = ± 1; x = ± 1, y = ± 4
5 x = 6, y = 2; x = −4, y = −3
6 x = 4, y = 1; x = −1, y = −4
7 x = 9, y = 2; x = 2 y = 9; x = −2, y = −9, x = −9, y = −2
8 x = 7, y = 1; x = 1, y = 7
9 x = 6, y = 3; x = −6, y = −3; x = 3, y = 6; x = −3, y = −6
10 x = 3, y = 1; x = 1.5, y = 2

Exercise 56 (p. 206)

1 (*a*) x < 2 or x > 3   (*b*) −2 < x < 4
  (*c*) x ⩽ 3 or x ⩾ 2
2 (*a*) x < 1 or x > 2   (*b*) x < 1 or x > 5
  (*c*) x ⩽ 1 or x ⩾ 4
3 (*a*) 0.5 < x < 3   (*b*) −2 < x < 0.25
  (*c*) x ⩽ −0.5 or x ⩾ 0.33 . . . (1 ÷ 3)
4 (*a*) x < 2 or 3 < x < 4   (*b*) x < −0.5 or 0.25 < x < 3
5 (*a*) 0 < x < 3   (*b*) 0 ⩽ x ⩽ 5
  (*c*) x > 5

Exercise 57 (p. 209)

1 $6a^9$ **2** a^{12} **3** $\frac{1}{8}a^8$
4 $2^{10} = 1024$ **5** $4^6 = 4096$ **6** $6a^8b^7$
7 $x^6y^{12}z^4$ **8** x^{2m} **9** a^{2m}
10 $a^{2p+q}b^p$ **11** a^{2n} **12** a^{m+n+4}
13 a^3 **14** $\frac{1}{2}x^3$ **15** $3a^5$
16 $2^6 = 64$ **17** $3^3 = 27$ **18** x^5
19 $-5a^3b$ **20** a^p **21** $4x^py^q$
22 a^7 **23** x **24** a^{n+1}
25 a^{2n} **26** a^6 **27** $\frac{8ab}{9}$
28 $\frac{a^2}{b^2}$ **29** $2^6 = 64$ **30** $3^6 = 729$
31 x^{10} **32** x^{10} **33** a^{16}
34 $4a^4b^6$ **35** $\frac{x^{12}y^9}{8}$ **36** a^{3p}
37 x^{4n} **38** $27a^{6p}$ **39** a^2

40 x^4 **41** x^8 **42** $3a^3$

43 a^2b **44** $\dfrac{x^3}{y^2}$ **45** $\dfrac{3a^4}{2b^3}$

46 x^n

Exercise 58 (p. 216)

1 (1) $\sqrt[4]{3}$ (2) $\sqrt{8}$ (3) $\sqrt[3]{a}$
   (4) $\sqrt{2}$ (5) $\sqrt[4]{3}$ (6) $\sqrt[4]{a^3}$

2 (1) $\sqrt[4]{1000}$ (2) $\sqrt[3]{16}$ (3) $\sqrt[4]{a^5}$
   (4) $10\sqrt{10}$ (5) $4\sqrt{2}$ (6) $10\sqrt[4]{10}$

3 (1) $2\sqrt[5]{2}$ (2) $4\sqrt[4]{2}$ (3) $\sqrt[4]{a^7}$
   (4) $\sqrt{a^7} = a^3\sqrt{a}$ (5) $\sqrt[5]{a^6} = a\sqrt[5]{a}$ (6) $\sqrt[5]{a}$

4 (1) 8   (2) 32   (3) 27
   (4) 1000   (5) 31.6   (6) $\frac{1}{2}$

5 $\frac{1}{9}, \frac{1}{3}, 1, 3, 3\sqrt{3}, 9, 9\sqrt{3}$

6 (1) 5.656   (2) 27   (3) $\frac{1}{10}$
   (4) $a^{\frac{17}{12}}$ (5) 2   (6) $a^{1.2}$

7 $\frac{1}{4}, \dfrac{3}{a^2}, \dfrac{1}{\sqrt{2}}, 3, 3a^2, \dfrac{1}{1000}$

8 (1) $\dfrac{1}{x^3}$ (2) $\dfrac{1}{x^{\frac{1}{2}}}$ (3) x^3

   (4) $\dfrac{1}{x^6}$ (5) x (6) $\frac{1}{2}x^{\frac{1}{3}}$

9 (1) 4   (2) 125   (3) 1000

   (4) $\dfrac{1}{5^6} = \dfrac{1}{156\ 25}$ (5) 16   (6) 3

10 (1) 4   (2) $\frac{27}{8}$ (3) $\frac{1}{4}$
   (4) $\frac{1}{6}$ (5) 8   (6) $\frac{1}{32}$

11 5.656

12 (1) $\sqrt[6]{a^5}$ (2) $100\sqrt{10} = 316.2$ (3) $\dfrac{1}{a^3}$

   (4) $\dfrac{1}{a^2}$ (5) x^{3n} (6) x^{2n}

13 (1) a^3 (2) 1   (3) 0
14 (1) 3.52×10^9 (2) 1.633×10^7 (3) 2.2×10^2
   (4) 3.7×10^3 (5) 2.52×10^{-7}

Exercise 59 (p. 224)

1 1, 3, 4, 2, 0, 5, 1, 3, 0, 2
2 (1) 0.6990, 1.6990, 2.6990, 4.6990
   (2) 0.6721, 2.6721, 4.6721
   (3) 1.7226, 0.7226, 2.7226
   (4) 2.9767, 0.9767, 4.9767
   (5) 0.7588, 1.9842, 3.8433
3 (1) 446.7, 44670, 44.67   (2) 87.70, 8770, 8.770
   (3) 4.714, 471.4, 471 400   (4) 2628, 5.229, 114.0

Exercise 60 (p. 227)

1	344.6	**2**	276.4	**3**	1397
4	5975	**5**	2.396	**6**	6.997
7	1.589	**8**	222.8	**9**	14.22
10	13.56	**11**	851.3	**12**	2650
13	3.137	**14**	728.5	**15**	2.172
16	104.6	**17**	1.564	**18**	1436
19	1.359	**20**	1.695	**21**	2.321
22	2.786	**23**	5.002	**24**	1.546

Exercise 61 (p. 229)

1 (1) 0.4470, $\overline{1}$.4470, $\overline{2}$.4470   (2) 0.6298, $\overline{1}$.6298, $\overline{3}$.6298
   (3) $\overline{3}$.9904, $\overline{4}$.9904, $\overline{1}$.9904   (4) $\overline{2}$.8097, $\overline{1}$.8097, $\overline{4}$.8097
2 (1) $\overline{2}$.7771   (2) $\overline{4}$.6749   (3) $\overline{3}$.9011
   (4) $\overline{5}$.9673   (5) $\overline{1}$.7538   (6) $\overline{2}$.9023
3 (1) 0.2159   (2) 0.007 453   (3) 0.030 70
   (4) 0.000 440 2 (5) 0.5940   (6) 2.482×10^{-8}

Exercise 62 (p. 231)

1 (1) $\overline{4}$.6037   (2) $\overline{2}$.7126
2 (1) $\overline{2}$.5926   (2) 0.8263   (3) $\overline{1}$.6597
   (4) 2.4814
3 (1) $\overline{1}$.7464   (2) $\overline{4}$.8368   (3) $\overline{3}$.8910
   (4) $\overline{1}$.3673   (5) $\overline{1}$.1958   (6) $\overline{4}$.7913
4 (1) $\overline{2}$.6856   (2) $\overline{1}$.071 55   (3) $\overline{1}$.7754
   (4) $\overline{1}$.1463   (5) $\overline{2}$.0254   (6) 0.5619
5 (1) $\overline{1}$.7399   (2) $\overline{1}$.7127   (3) $\overline{2}$.7726
   (4) $\overline{2}$.5598   (5) $\overline{1}$.7266   (6) $\overline{3}$.8973

Exercise 63 (p. 234)

1	15.42	**2**	0.3285	**3**	0.015 29
4	5.699	**5**	0.6116	**6**	0.032 39
7	0.049 03	**8**	0.1600	**9**	85.23
10	0.8414	**11**	0.1226	**12**	1.197
13	0.071 15	**14**	1.826	**15**	1.457
16	3.558	**17**	5.471	**18**	0.1014
19	0.1429	**20**	9.399	**21**	0.3609
22	0.031 29	**23**	1.022	**24**	62.84
25	1.663	**26**	3.940	**27**	9.527
28	896.6	**29**	102.2	**30**	516.4
31	2.89	**32**	1.035	**33**	3.98
34	0.795	**35**	9.81	**36**	(*a*) 3; (*b*) 3
37	530.7	**38**	4.923 mm	**39**	(1) 1.5261
40	1200	**41**	0.8268		(2) $\overline{3}$.2193

Exercise 64 (p. 241)

1 (*a*) $100a:b$ (*b*) $\dfrac{3600p}{60q + r}$ **2** $36 \times 10^5 : 1$

3 (1) $\frac{3}{5}$ (2) $25:64$ (3) $5:12$

4 (*a*) 35 (*b*) $\dfrac{bx}{a}$ (*c*) $\dfrac{a}{x}$

5 $\dfrac{c(b + c)}{a}$ **6** $\dfrac{pA}{p + q} , \dfrac{qA}{p + q}$

7 $\dfrac{ax}{x + y}$ kg, $\dfrac{ay}{x + y}$ kg   **8** $\dfrac{bc}{a}$ **9** 17

10 (1) $b\sqrt{ac}$ (2) $4ab$ (3) $(a + b)\sqrt{ab}$

11 7 **12** $\frac{7}{3}$ **13** (1) $\frac{7}{3}$ (2) $\frac{3}{7}$ (3) $\frac{5}{4}$

14 (*a*) 450 (*b*) $\dfrac{6a}{5b}$ (*c*) $\dfrac{4a}{3b}$

15 $x = \sqrt[3]{a^2b}, y = \sqrt[3]{ab^2}$

Exercise 65 (p. 249)

1 (*a*) No. The runner's rate differs at various parts of the race.
    (*b*) Yes   (*c*) No   (*d*) Yes
    (*e*) No. The connecting law is as explained in Section 221.

2 $\frac{8}{7}; 45\frac{5}{7}$ **3** $37.5; 3.36$

4 $y = \frac{33}{7}x; 4\frac{5}{11}$ **5** $s = 3.2t; 8.96$ m

6 $E = \frac{8}{15}W; 3.73$ **7** $y = \frac{1}{2}x + 6$

8 $E = 0.06W + 0.2$ **9** $F = 0.42W - 0.6$

Exercise 66 (p. 259)

1 $y = \frac{8}{9}x^2$; $64\frac{2}{9}$

2 $y = \frac{1}{4}x^3$; $\frac{27}{4}$

3 y values 12, 3; x value 6

4 $y = \frac{7}{4}\sqrt{x}$; 8.75

5 384; $2\sqrt[3]{4}$

6 $y = \dfrac{30}{x}$; 1.5

7 x values 2, 15; y value 10

8 $F = \dfrac{1920}{d^2}$

9 4 s

10 10 m

11 102 mm

12 1.8

13 50 s

14 $y = 8x^{1.4}$

Exercise 67 (p. 261)

1 (*a*) $y \propto xz$ $\therefore y = kxz$

(*b*) $y \propto \dfrac{x}{z^2}$ $\therefore y = \dfrac{kx}{z^2}$

(*c*) $y \propto \dfrac{\sqrt{x}}{z}$ $\therefore y = \dfrac{k\sqrt{x}}{z}$

(*d*) $V \propto hr^2$ $\therefore y = khr^2$

(*e*) $W \propto \dfrac{bd^2}{l}$ $\therefore W = k.\dfrac{bd^2}{l}$

(*f*) $y \propto \dfrac{x^2}{\sqrt[3]{z}}$ $\therefore y = k.\dfrac{x^2}{\sqrt[3]{z}}$

2 $y = \dfrac{25}{4}\dfrac{x}{z}$, 15

3 $y = 3xz^2$; 8

4 $10\frac{5}{12}$

5 8 t

6 1

7 $H = \dfrac{0.24IE^2}{R}$ (1) 24 000 units   (2) 150 s

8 4.1% increase.

Exercise 68 (p. 268)

1 $y = 6x^2 + 3$

2 $y = 0.42x^2 + 7.5$

3 $R = 0.01V^2 + 5$

4 $H = 1.08V^3 + 420$

5 $y = 0.54 + 0.24x^2$

6 $R = \dfrac{1200}{V} + 50$

7 $y = 0.25x^3$

8 $y = 3x^{\frac{1}{3}}$

9 $Q = 5.7H^{0.36}$

10 $y = 3.65x^{\frac{1}{4}}$

Exercise 69 (p. 275)

1   (1) $\sqrt{150}$ (2) $\sqrt{288}$ (3) $\sqrt{500}$
    (4) $\sqrt{208}$ (5) $\sqrt{4a^2b}$ (6) $\sqrt{9x^4y}$

2   (1) $20\sqrt{2}$ (2) $8\sqrt{5}$ (3) $6\sqrt{3}$
    (4) $50\sqrt{2}$ (5) $5\sqrt{15}$ (6) $60\sqrt{2}$
    (7) $2ab\sqrt{6a}$ (8) $3x^2y\sqrt{2yz}$ (9) $5abc\sqrt{3abc}$
  (10) $10p\sqrt{10q}$ (11) $a(a-b)\sqrt{b(a-b)}$
  (12) $5y(x-2y)\sqrt{xy}$

3   (1) $3\sqrt{5}$ (2) $7\sqrt{2}$ (3) $16\sqrt{3}$
    (4) $10\sqrt{10}$ (5) $4-\sqrt{2}$ (6) $7\sqrt{2}+\sqrt{14}$
    (7) $(a+b)\sqrt{a-b}$ (8) $3(x-2y)\sqrt{2(x+2y)}$

4   (1) $2\sqrt{2}-1$ (2) $2-3\sqrt{6}$ (3) $3-2\sqrt{2}$
    (4) $23+4\sqrt{15}$ (5) $57-10\sqrt{14}$ (6) 2
    (7) 38 (8) $301+20\sqrt{3}$ (9) $a-25$
  (10) 21

5   (1) $\dfrac{2\sqrt{3}}{3}$ (2) $\dfrac{\sqrt{5}}{5}$ (3) $\dfrac{3\sqrt{7}}{14}$

    (4) $\dfrac{\sqrt{10}}{30}$ (5) $\dfrac{\sqrt{2}}{8}$ (6) $\dfrac{6\sqrt{5}}{5}$

    (7) $\dfrac{\sqrt{6}}{3}$ (8) $\dfrac{2\sqrt{35}}{7}$ (9) $\dfrac{\sqrt{5}+1}{4}$

  (10) $\dfrac{3(\sqrt{7}-\sqrt{2})}{5}$ (11) $\dfrac{3(2\sqrt{3}+\sqrt{2})}{10}$

  (12) $5(2\sqrt{7}+3\sqrt{3})$ (13) $5+2\sqrt{6}$

  (14) $6-\sqrt{35}$ (15) $\dfrac{21+\sqrt{7}}{62}$

  (16 $2\sqrt{2}$ (17) $-\sqrt{3}$

  (18) $\dfrac{4(6+\sqrt{5})}{31}$

Exercise 70 (p. 281)

1   (1) $12.5, 15, 17.5$ (2) $0, -4, -8$
    (3) $(a-3b), (a-5b), (a-7b)$
    (4) $6.6, 7.9, 9.2$
    (5) $(x+2y), (x+3y), (x+4y)$

2 $12, 16.5$ **3** $4+4p$

4 $x+3n-1$ **5** 3.48

6 29 **7** 2 **8** 19

9   (*a*) 217.5 (*b*) 16 (*c*) 0 (*d*) $364\frac{1}{2}$

10 12 **11** 5 or 8 **12** $120; 1860$

13   £40.05; £5343.75   **14**   £34

Exercise 71 (p. 285)

1 (a) 62.5, 156.25, 390.625   (b) $\frac{1}{4}, \frac{1}{16}, \frac{1}{64}$
  (c) $-54, 81, -\frac{243}{2}$
  (d) $3 \times 10^{-4}, 3 \times 10^{-5}, 3 \times 10^{-6}$
  (e 0.010 125, 0.001 518 75, 0.000 227 8 (approx.)
2 320   3 $\frac{128}{234}$
4 1.610 51   5 0.001 215
6 (a) $ar^{2n-1}, ar^{2n}, -ar^{2n-1}, ar^{2n}$
7 1.05   8 1.04
9 (1) 3.873; (2) 3.888 (both approx.)
10 7, 9.8   11 £1464.10   12 £162 901.25
13 1.04   14 10, 20, 40

Exercise 72 (p. 288)

1 (a) 94.5   (b) 19.93   (c) $\frac{21}{64}$
2 $\frac{315}{32}$ 3 $\frac{133}{243}$ 4 16.32
5 216.8   6 765   7 27; $40\frac{40}{81}$
8 93.8 approx.

Exercise 73 (p. 293)

1 (a) $\frac{2}{3}$ (b) $\frac{1}{3}$ (c) 1
2 (a) $\frac{10}{99}$ (b) $\frac{6}{99}$ (c) $\frac{16}{99}$
  Series (c) is the sum of (a) and (b). It represents 0.161 616 . . . ,
  a recurring decimal.
3 $\frac{2}{3}$. This series equals the series (b) in question (1) increased by
  unity and then diminished by series (a).
4 In $\frac{3}{2}\{1 - (\frac{1}{3})^n\}$, $(\frac{1}{3})^n \to$ o as n $\to \infty$. $\therefore$ the limit is $\frac{3}{2}$.
5 $\dfrac{b}{b - a}\left\{1 - \left(\dfrac{a}{b}\right)^n\right\}$; limit = $\dfrac{b}{b - a}$.
6 (a) $\frac{16}{5}$ (b) $\frac{25}{6}$ (c) $\frac{27}{5}$
7 (a) 26   (b) $2(\sqrt{2} + 1)$
8 $\frac{4}{5}$ 9 180 m   10 £75 000

Exercise 74 (p. 298)

Most of the answers are approximate.

1 £1180   2 £5500   3 £486
4 £2430   5 £3431   6 £5000
7 £1667   8 £240 000

LOGARITHMS

	0	1	2	3	4	5	6	7	8	9	1	2	3	4	5	6	7	8	9
10	0000	0043	0086	0128	0170	0212	0253	0294	0334	0374	4	8	12	17	21	25	29	33	37
11	0414	0453	0492	0531	0569	0607	0645	0682	0719	0755	4	8	11	15	19	23	26	30	34
12	0792	0828	0864	0899	0934	0969	1004	1038	1072	1106	3	7	10	14	17	21	24	28	31
13	1139	1173	1206	1239	1271	1303	1335	1367	1399	1430	3	6	10	13	16	19	23	26	29
14	1461	1492	1523	1553	1584	1614	1644	1673	1703	1732	3	6	9	12	15	18	21	24	27
15	1761	1790	1818	1847	1875	1903	1931	1959	1987	2014	3	6	8	11	14	17	20	22	25
16	2041	2068	2095	2122	2148	2175	2201	2227	2253	2279	3	5	8	11	13	16	18	21	24
17	2304	2330	2355	2380	2405	2430	2455	2480	2504	2529	2	5	7	10	12	15	17	20	22
18	2553	2577	2601	2625	2648	2672	2695	2718	2742	2765	2	5	7	9	12	14	16	19	21
19	2788	2810	2833	2856	2878	2900	2923	2945	2967	2989	2	4	7	9	11	13	16	18	20
20	3010	3032	3054	3075	3096	3118	3139	3160	3181	3201	2	4	6	8	11	13	15	17	19
21	3222	3243	3263	3284	3304	3324	3345	3365	3385	3404	2	4	6	8	10	12	14	16	18
22	3424	3444	3464	3483	3502	3522	3541	3560	3579	3598	2	4	6	8	10	12	14	15	17
23	3617	3636	3655	3674	3692	3711	3729	3747	3766	3784	2	4	6	7	9	11	13	15	17
24	3802	3820	3838	3856	3874	3892	3909	3927	3945	3962	2	4	5	7	9	11	12	14	16
25	3979	3997	4014	4031	4048	4065	4082	4099	4116	4133	2	3	5	7	9	10	12	14	15
26	4150	4166	4183	4200	4216	4232	4249	4265	4281	4298	2	3	5	7	8	10	11	13	15
27	4314	4330	4346	4362	4378	4393	4409	4425	4440	4456	2	3	5	6	8	9	11	13	14
28	4472	4487	4502	4518	4533	4548	4564	4579	4594	4609	2	3	5	6	8	9	11	12	14
29	4624	4639	4654	4669	4683	4698	4713	4728	4742	4757	1	3	4	6	7	9	10	12	13
30	4771	4786	4800	4814	4829	4843	4857	4871	4886	4900	1	3	4	6	7	9	10	11	13
31	4914	4928	4942	4955	4969	4983	4997	5011	5024	5038	1	3	4	5	7	8	10	11	12
32	5051	5065	5079	5092	5105	5119	5132	5145	5159	5172	1	3	4	5	7	8	9	11	12
33	5185	5198	5211	5224	5237	5250	5263	5276	5289	5302	1	3	4	5	6	8	9	10	12
34	5315	5328	5340	5353	5366	5378	5391	5403	5416	5428	1	3	4	5	6	8	9	10	11
35	5441	5453	5465	5478	5490	5502	5514	5527	5539	5551	1	2	4	5	6	7	9	10	11
36	5563	5575	5587	5599	5611	5623	5635	5647	5658	5670	1	2	4	5	6	7	8	10	11
37	5682	5694	5705	5717	5729	5740	5752	5763	5775	5786	1	2	3	5	6	7	8	9	10
38	5798	5809	5821	5832	5843	5855	5866	5877	5888	5899	1	2	3	5	6	7	8	9	10
39	5911	5922	5933	5944	5955	5966	5977	5988	5999	6010	1	2	3	4	5	7	8	9	10
40	6021	6031	6042	6053	6064	6075	6085	6096	6107	6117	1	2	3	4	5	6	7	9	10
41	6128	6138	6149	6160	6170	6180	6191	6201	6212	6222	1	2	3	4	5	6	7	8	9
42	6232	6243	6253	6263	6274	6284	6294	6304	6314	6325	1	2	3	4	5	6	7	8	9
43	6335	6345	6355	6365	6375	6385	6395	6405	6415	6425	1	2	3	4	5	6	7	8	9
44	6435	6444	6454	6464	6474	6484	6493	6503	6513	6522	1	2	3	4	5	6	7	8	9
45	6532	6542	6551	6561	6571	6580	6590	6599	6609	6618	1	2	3	4	5	6	7	8	9
46	6628	6637	6646	6656	6665	6675	6684	6693	6702	6712	1	2	3	4	5	6	7	7	8
47	6721	6730	6739	6749	6758	6767	6776	6785	6794	6803	1	2	3	4	5	5	6	7	8
48	6812	6821	6830	6839	6848	6857	6866	6875	6884	6893	1	2	3	4	4	5	6	7	8
49	6902	6911	6920	6928	6937	6946	6955	6964	6972	6981	1	2	3	4	4	5	6	7	8
50	6990	6998	7007	7016	7024	7033	7042	7050	7059	7067	1	2	3	3	4	5	6	7	8
51	7076	7084	7093	7101	7110	7118	7126	7135	7143	7152	1	2	3	3	4	5	6	7	8
52	7160	7168	7177	7185	7193	7202	7210	7218	7226	7235	1	2	2	3	4	5	6	7	7
53	7243	7251	7259	7267	7275	7284	7292	7300	7308	7316	1	2	2	3	4	5	6	6	7
54	7324	7332	7340	7348	7356	7364	7372	7380	7388	7396	1	2	2	3	4	5	6	6	7
	0	1	2	3	4	5	6	7	8	9	1	2	3	4	5	6	7	8	9

LOGARITHMS

	0	1	2	3	4	5	6	7	8	9	1	2	3	4	5	6	7	8	9
55	7404	7412	7419	7427	7435	7443	7451	7459	7466	7474	1	2	2	3	4	5	5	6	7
56	7482	7490	7497	7505	7513	7520	7528	7536	7543	7551	1	2	2	3	4	5	5	6	7
57	7559	7566	7574	7582	7589	7597	7604	7612	7619	7627	1	2	2	3	4	5	5	6	7
58	7634	7642	7649	7657	7664	7672	7679	7686	7694	7701	1	1	2	3	4	4	5	6	7
59	7709	7716	7723	7731	7738	7745	7752	7760	7767	7774	1	1	2	3	4	4	5	6	7
60	7782	7789	7796	7803	7810	7818	7825	7832	7839	7846	1	1	2	3	4	4	5	6	6
61	7853	7860	7868	7875	7882	7889	7896	7903	7910	7917	1	1	2	3	4	4	5	6	6
62	7924	7931	7938	7945	7952	7959	7966	7973	7980	7987	1	1	2	3	3	4	5	6	6
63	7993	8000	8007	8014	8021	8028	8035	8041	8048	8055	1	1	2	3	3	4	5	6	6
64	8062	8069	8075	8082	8089	8096	8102	8109	8116	8122	1	1	2	3	3	4	5	5	6
65	8129	8136	8142	8149	8156	8162	8169	8176	8182	8189	1	1	2	3	3	4	5	5	6
66	8195	8202	8209	8215	8222	8228	8235	8241	8248	8254	1	1	2	3	3	4	5	5	6
67	8261	8267	8274	8280	8287	8293	8299	8306	8312	8319	1	1	2	3	3	4	4	5	6
68	8325	8331	8338	8344	8351	8357	8363	8370	8376	8382	1	1	2	3	3	4	4	5	6
69	8388	8395	8401	8407	8414	8420	8426	8432	8439	8445	1	1	2	3	3	4	4	5	6
70	8451	8457	8463	8470	8476	8482	8488	8494	8500	8506	1	1	2	2	3	4	4	5	6
71	8513	8519	8525	8531	8537	8543	8549	8555	8561	8567	1	1	2	2	3	4	4	5	5
72	8573	8579	8585	8591	8597	8603	8609	8615	8621	8627	1	1	2	2	3	4	4	5	5
73	8633	8639	8645	8651	8657	8663	8669	8675	8681	8686	1	1	2	2	3	4	4	5	5
74	8692	8698	8704	8710	8716	8722	8727	8733	8739	8745	1	1	2	2	3	4	4	5	5
75	8751	8756	8762	8768	8774	8779	8785	8791	8797	8802	1	1	2	2	3	3	4	5	5
76	8808	8814	8820	8825	8831	8837	8842	8848	8854	8859	1	1	2	2	3	3	4	5	5
77	8865	8871	8876	8882	8887	8893	8899	8904	8910	8915	1	1	2	2	3	3	4	4	5
78	8921	8927	8932	8938	8943	8949	8954	8960	8965	8971	1	1	2	2	3	3	4	4	5
79	8976	8982	8987	8993	8998	9004	9009	9015	9020	9025	1	1	2	2	3	3	4	4	5
80	9031	9036	9042	9047	9053	9058	9063	9069	9074	9079	1	1	2	2	3	3	4	4	5
81	9085	9090	9096	9101	9106	9112	9117	9122	9128	9133	1	1	2	2	3	3	4	4	5
82	9138	9143	9149	9154	9159	9165	9170	9175	9180	9186	1	1	2	2	3	3	4	4	5
83	9191	9196	9201	9206	9212	9217	9222	9227	9232	9238	1	1	2	2	3	3	4	4	5
84	9243	9248	9253	9258	9263	9269	9274	9279	9284	9289	1	1	2	2	3	3	4	4	5
85	9294	9299	9304	9309	9315	9320	9325	9330	9335	9340	1	1	2	2	3	3	4	4	5
86	9345	9350	9355	9360	9365	9370	9375	9380	9385	9390	1	1	2	2	3	3	4	4	5
87	9395	9400	9405	9410	9415	9420	9425	9430	9435	9440	0	1	1	2	2	3	3	4	4
88	9445	9450	9455	9460	9465	9469	9474	9479	9484	9489	0	1	1	2	2	3	3	4	4
89	9494	9499	9504	9509	9513	9518	9523	9528	9533	9538	0	1	1	2	2	3	3	4	4
90	9542	9547	9552	9557	9562	9566	9571	9576	9581	9586	0	1	1	2	2	3	3	4	4
91	9590	9595	9600	9605	9609	9614	9619	9624	9628	9633	0	1	1	2	2	3	3	4	4
92	9638	9643	9647	9652	9657	9661	9666	9671	9675	9680	0	1	1	2	2	3	3	4	4
93	9685	9689	9694	9699	9703	9708	9713	9717	9722	9727	0	1	1	2	2	3	3	4	4
94	9731	9736	9741	9745	9750	9754	9759	9764	9768	9773	0	1	1	2	2	3	3	4	4
95	9777	9782	9786	9791	9795	9800	9805	9809	9814	9818	0	1	1	2	2	3	3	4	4
96	9823	9827	9832	9836	9841	9845	9850	9854	9859	9863	0	1	1	2	2	3	3	4	4
97	9868	9872	9877	9881	9886	9890	9894	9899	9903	9908	0	1	1	2	2	3	3	4	4
98	9912	9917	9921	9926	9930	9934	9939	9943	9948	9952	0	1	1	2	2	3	3	4	4
99	9956	9961	9965	9969	9974	9978	9983	9987	9991	9996	0	1	1	2	2	3	3	4	4
	0	1	2	3	4	5	6	7	8	9	1	2	3	4	5	6	7	8	9

ANTI-LOGARITHMS

	0	1	2	3	4	5	6	7	8	9	1	2	3	4	5	6	7	8	9
.00	1000	1002	1005	1007	1009	1012	1014	1016	1019	1021	0	0	1	1	1	1	2	2	2
.01	1023	1026	1028	1030	1033	1035	1038	1040	1042	1045	0	0	1	1	1	1	2	2	2
.02	1047	1050	1052	1054	1057	1059	1062	1064	1067	1069	0	0	1	1	1	1	2	2	2
.03	1072	1074	1076	1079	1081	1084	1086	1089	1091	1094	0	0	1	1	1	1	2	2	2
.04	1096	1099	1102	1104	1107	1109	1112	1114	1117	1119	0	1	1	1	1	2	2	2	2
.05	1122	1125	1127	1130	1132	1135	1138	1140	1143	1146	0	1	1	1	1	2	2	2	2
.06	1148	1151	1153	1156	1159	1161	1164	1167	1169	1172	0	1	1	1	1	2	2	2	2
.07	1175	1178	1180	1183	1186	1189	1191	1194	1197	1199	0	1	1	1	1	2	2	2	2
.08	1202	1205	1208	1211	1213	1216	1219	1222	1225	1227	0	1	1	1	1	2	2	2	3
.09	1230	1233	1236	1239	1242	1245	1247	1250	1253	1256	0	1	1	1	1	2	2	2	3
.10	1259	1262	1265	1268	1271	1274	1276	1279	1282	1285	0	1	1	1	1	2	2	2	3
.11	1288	1291	1294	1297	1300	1303	1306	1309	1312	1315	0	1	1	2	2	2	2	2	3
.12	1318	1321	1324	1327	1330	1334	1337	1340	1343	1346	0	1	1	2	2	2	2	3	3
.13	1349	1352	1355	1358	1361	1365	1368	1371	1374	1377	0	1	1	1	2	2	2	3	3
.14	1380	1384	1387	1390	1393	1396	1400	1403	1406	1409	0	1	1	1	2	2	2	3	3
.15	1413	1416	1419	1422	1426	1429	1432	1435	1439	1442	0	1	1	1	2	2	2	3	3
.16	1445	1449	1452	1455	1459	1462	1466	1469	1472	1476	0	1	1	1	2	2	2	3	3
.17	1479	1483	1486	1489	1493	1496	1500	1503	1507	1510	0	1	1	1	2	2	2	3	3
.18	1514	1517	1521	1524	1528	1531	1535	1538	1542	1545	0	1	1	1	2	2	2	3	3
.19	1549	1552	1556	1560	1563	1567	1570	1574	1578	1581	0	1	1	1	2	2	3	3	3
.20	1585	1589	1592	1596	1600	1603	1607	1611	1614	1618	0	1	1	1	2	2	3	3	3
.21	1622	1626	1629	1633	1637	1641	1644	1648	1652	1656	0	1	1	2	2	2	3	3	3
.22	1660	1663	1667	1671	1675	1679	1683	1687	1690	1694	0	1	1	2	2	2	3	3	3
.23	1698	1702	1706	1710	1714	1718	1722	1726	1730	1734	0	1	1	2	2	2	3	3	4
.24	1738	1742	1746	1750	1754	1758	1762	1766	1770	1774	0	1	1	2	2	2	3	3	4
.25	1778	1782	1786	1791	1795	1799	1803	1807	1811	1816	0	1	1	2	2	3	3	3	4
.26	1820	1824	1828	1832	1837	1841	1845	1849	1854	1858	0	1	1	2	2	3	3	3	4
.27	1862	1866	1871	1875	1879	1884	1888	1892	1897	1901	0	1	1	2	2	3	3	3	4
.28	1905	1910	1914	1919	1923	1928	1932	1936	1941	1945	0	1	1	2	2	3	3	4	4
.29	1950	1954	1959	1963	1968	1972	1977	1982	1986	1991	0	1	1	2	2	3	3	4	4
.30	1995	2000	2004	2009	2014	2018	2023	2028	2032	2037	0	1	1	2	2	3	3	4	4
.31	2042	2046	2051	2056	2061	2065	2070	2075	2080	2084	0	1	1	2	2	3	3	4	4
.32	2089	2094	2099	2104	2109	2113	2118	2123	2128	2133	0	1	1	2	2	3	3	4	4
.33	2138	2143	2148	2153	2158	2163	2168	2173	2178	2183	0	1	1	2	2	3	3	4	4
.34	2188	2193	2198	2203	2208	2213	2218	2223	2228	2234	1	1	2	2	3	3	4	4	5
.35	2239	2244	2249	2254	2259	2265	2270	2275	2280	2286	1	1	2	2	3	3	4	4	5
.36	2291	2296	2301	2307	2312	2317	2323	2328	2333	2339	1	1	2	2	3	3	4	4	5
.37	2344	2350	2355	2360	2366	2371	2377	2382	2388	2393	1	1	2	2	3	3	4	4	5
.38	2399	2404	2410	2415	2421	2427	2432	2438	2443	2449	1	1	2	2	3	3	4	4	5
.39	2455	2460	2466	2472	2477	2483	2489	2495	2500	2506	1	1	2	2	3	3	4	5	5
.40	2512	2518	2523	2529	2535	2541	2547	2553	2559	2564	1	1	2	2	3	3	4	5	5
.41	2570	2576	2582	2588	2594	2600	2606	2612	2618	2624	1	1	2	2	3	4	4	5	5
.42	2630	2636	2642	2648	2655	2661	2667	2673	2679	2685	1	1	2	2	3	4	4	5	6
.43	2692	2698	2704	2710	2716	2723	2729	2735	2742	2748	1	1	2	2	3	4	4	5	6
.44	2754	2761	2767	2773	2780	2786	2793	2799	2805	2812	1	1	2	3	3	4	4	5	6
.45	2818	2825	2831	2838	2844	2851	2858	2864	2871	2877	1	1	2	3	3	4	5	5	6
.46	2884	2891	2897	2904	2911	2917	2924	2931	2938	2944	1	1	2	3	3	4	5	5	6
.47	2951	2958	2965	2972	2979	2985	2992	2999	3006	3013	1	1	2	3	3	4	5	6	6
.48	3020	3027	3034	3041	3048	3055	3062	3069	3076	3083	1	1	2	3	4	4	5	6	6
.49	3090	3097	3105	3112	3119	3126	3133	3141	3148	3155	1	1	2	3	4	4	5	6	7
	0	1	2	3	4	5	6	7	8	9	1	2	3	4	5	6	7	8	9

ANTI-LOGARITHMS

	0	1	2	3	4	5	6	7	8	9	1	2	3	4	5	6	7	8	9
.50	3162	3170	3177	3184	3192	3199	3206	3214	3221	3228	1	1	2	3	4	4	5	6	7
.51	3236	3243	3251	3258	3266	3273	3281	3289	3296	3304	1	2	2	3	4	5	5	6	7
.52	3311	3319	3327	3334	3342	3350	3357	3365	3373	3381	1	2	2	3	4	5	5	6	7
.53	3388	3396	3404	3412	3420	3428	3436	3443	3451	3459	1	2	2	3	4	5	6	6	7
.54	3467	3475	3483	3491	3499	3508	3516	3524	3532	3540	1	2	2	3	4	5	6	6	7
.55	3548	3556	3565	3573	3581	3589	3597	3606	3614	3622	1	2	2	3	4	5	6	7	7
.56	3631	3639	3648	3656	3664	3673	3681	3690	3698	3707	1	2	3	3	4	5	6	7	8
.57	3715	3724	3733	3741	3750	3758	3767	3776	3784	3793	1	2	3	3	4	5	6	7	8
.58	3802	3811	3819	3828	3837	3846	3855	3864	3873	3882	1	2	3	4	4	5	6	7	8
.59	3890	3899	3908	3917	3926	3936	3945	3954	3963	3972	1	2	3	4	5	5	6	7	8
.60	3981	3990	3999	4009	4018	4027	4036	4046	4055	4064	1	2	3	4	5	5	7	7	8
.61	4074	4083	4093	4102	4111	4121	4130	4140	4150	4159	1	2	3	4	5	6	7	8	9
.62	4169	4178	4188	4198	4207	4217	4227	4236	4246	4256	1	2	3	4	5	6	7	8	9
.63	4266	4276	4285	4295	4305	4315	4325	4335	4345	4355	1	2	3	4	5	6	7	8	9
.64	4365	4375	4385	4395	4406	4416	4426	4436	4446	4457	1	2	3	4	5	6	7	8	9
.65	4467	4477	4487	4498	4508	4519	4529	4539	4550	4560	1	2	3	4	5	6	7	8	9
.66	4571	4581	4592	4603	4613	4624	4634	4645	4656	4667	1	2	3	4	5	6	7	8	10
.67	4677	4688	4699	4710	4721	4732	4742	4753	4764	4775	1	2	3	4	5	7	8	9	10
.68	4786	4797	4808	4819	4831	4842	4853	4864	4875	4887	1	2	3	4	6	7	8	9	10
.69	4898	4909	4920	4932	4943	4955	4966	4977	4989	5000	1	2	3	5	6	7	8	9	10
.70	5012	5023	5035	5047	5058	5070	5082	5093	5105	5117	1	2	4	5	6	7	8	9	11
.71	5129	5140	5152	5164	5176	5188	5200	5212	5224	5236	1	2	4	5	6	7	8	10	11
.72	5248	5260	5272	5284	5297	5309	5321	5333	5346	5358	1	2	4	5	6	7	9	10	11
.73	5370	5383	5395	5408	5420	5433	5445	5458	5470	5483	1	3	4	5	6	8	9	10	11
.74	5495	5508	5521	5534	5546	5559	5572	5585	5598	5610	1	3	4	5	6	8	9	10	12
.75	5623	5636	5649	5662	5675	5689	5702	5715	5728	5741	1	3	4	5	7	8	9	10	12
.76	5754	5768	5781	5794	5808	5821	5834	5848	5861	5875	1	3	4	5	7	8	9	11	12
.77	5888	5902	5916	5929	5943	5957	5970	5984	5998	6012	1	3	4	6	7	8	10	11	12
.78	6026	6039	6053	6067	6081	6095	6109	6124	6138	6152	1	3	4	6	7	8	10	11	13
.79	6166	6180	6194	6209	6223	6237	6252	6266	6281	6295	1	3	4	6	7	9	10	12	13
.80	6310	6324	6339	6353	6368	6383	6397	6412	6427	6442	1	3	4	6	7	9	10	12	13
.81	6457	6471	6486	6501	6516	6531	6546	6561	6577	6592	2	3	5	6	8	9	11	12	14
.82	6607	6622	6637	6653	6668	6683	6699	6714	6730	6745	2	3	5	6	8	9	11	12	14
.83	6761	6776	6792	6808	6823	6839	6855	6871	6887	6902	2	3	5	6	8	9	11	13	14
.84	6918	6934	6950	6966	6982	6998	7015	7031	7047	7063	2	3	5	6	8	10	11	13	14
.85	7079	7096	7112	7129	7145	7161	7178	7194	7211	7228	2	3	5	7	8	10	12	13	15
.86	7244	7261	7278	7295	7311	7328	7345	7362	7379	7396	2	3	5	7	8	10	12	14	15
.87	7413	7430	7447	7464	7482	7499	7516	7534	7551	7568	2	3	5	7	9	10	12	14	16
.88	7586	7603	7621	7638	7656	7674	7691	7709	7727	7745	2	4	5	7	9	11	12	14	16
.89	7762	7780	7798	7816	7834	7852	7870	7889	7907	7925	2	4	5	7	9	11	13	14	16
.90	7943	7962	7980	7998	8017	8035	8054	8072	8091	8110	2	4	6	7	9	11	13	15	17
.91	8128	8147	8166	8185	8204	8222	8241	8260	8279	8299	2	4	6	8	10	11	13	15	17
.92	8318	8337	8356	8375	8395	8414	8433	8453	8472	8492	2	4	6	8	10	12	14	15	17
.93	8511	8531	8551	8570	8590	8610	8630	8650	8670	8690	2	4	6	8	10	12	14	16	18
.94	8710	8730	8750	8770	8790	8810	8831	8851	8872	8892	2	4	6	8	10	12	14	16	18
.95	8913	8933	8954	8974	8995	9016	9036	9057	9078	9099	2	4	6	8	10	12	14	17	19
.96	9120	9141	9162	9183	9204	9226	9247	9268	9290	9311	2	4	6	9	11	13	15	17	19
.97	9333	9354	9376	9397	9419	9441	9462	9484	9506	9528	2	4	7	9	11	13	15	17	20
.98	9550	9572	9594	9616	9638	9661	9683	9705	9727	9750	2	4	7	9	11	13	16	18	20
.99	9772	9795	9817	9840	9863	9886	9908	9931	9954	9977	2	5	7	9	11	14	16	18	21
	0	1	2	3	4	5	6	7	8	9	1	2	3	4	5	6	7	8	9

OTHER TITLES AVAILABLE
IN TEACH YOURSELF

P. ABBOTT (REVISED BY MICHAEL WARDLE)

☐ 0 340 54913 0   Trigonometry   £4.99

☐ 0 340 54912 2   Calculus   £6.99

L.C. PASCOE

☐ 0 340 28766 7   Mathematics   £4.99

All these books are available at your local bookshop or newsagent, or can be ordered direct from the publisher. Just tick the titles you want and fill in the form below.

Prices and availability subject to change without notice.

HODDER AND STOUGHTON PAPERBACKS, P O Box 11, Falmouth, Cornwall.

Please send cheque or postal order for the value of the book, and add the following for postage and packing:

U.K. – 80p for one book, and 20p for each additional book ordered up to a £2.00 maximum.

B.F.P.O. – 80p for the first book, and 20p for each additional book. OVERSEAS INCLUDING EIRE – £1.50 for the first book, plus £1.00 for the second book, and 30p for each additional book ordered.
OR Please debit this amount from my Access/Visa Card (delete as appropriate).

Card Number ☐☐☐☐☐☐☐☐☐☐☐☐☐☐☐☐

AMOUNT £ ....

EXPIRY DATE ....

SIGNED ....

NAME ....

ADDRESS ....